Liberty Hyde Bailey

First Lessons with Plants

being an abridgement

Liberty Hyde Bailey

First Lessons with Plants
being an abridgement

ISBN/EAN: 9783337363550

Printed in Europe, USA, Canada, Australia, Japan

Cover: Foto ©berggeist007 / pixelio.de

More available books at **www.hansebooks.com**

FIRST LESSONS WITH PLANTS

BEING AN ABRIDGEMENT OF

"LESSONS WITH PLANTS:

SUGGESTIONS FOR SEEING AND INTERPRETING SOME OF
THE COMMON FORMS OF VEGETATION"

BY

L. H. BAILEY

With delineations from nature by
W. S. HOLDSWORTH
Assistant Professor of Drawing in the Agricultural
College of Michigan

THIRD EDITION

New York
THE MACMILLAN COMPANY
LONDON: MACMILLAN & CO., LTD.
1906

Mount Pleasant Press
J. HORACE McFARLAND COMPANY
HARRISBURG · PENNSYLVANIA

PREFACE

These simple lessons are designed to awaken an interest in plants and in nature rather than to teach botany. They are suggestions to the teacher who desires to introduce nature-study into the school. A somewhat full discussion of the author's opinions respecting the methods of presenting nature-study by means of plant-subjects, is given in the book of which this is an abridgement. It is desired to emphasize the importance of making nature-study objects the subjects of writing and drawing in schools in which composition and drawing are taught. The first essential to the writing of compositions is that the pupil have something to say which is drawn from experience and observation. Live and emphatic ideas are more important than drill in modes of expression. Fill the pupil with his subject, and writing comes easy, particularly if he is taught that good English demands that he go no farther with his subject than to express what he himself feels. The writing and the drawing should not be intended, primarily, as examinations in the nature-study, but as regular exercises in the customary work of composition and drawing.

(v)

Teachers sometimes like to take up the plant as an entirety, before discussing its parts. Familiar plants may be brought before the class, and the different parts pointed out,—as stems, roots, leaves, flowers. This is desirable with children, but its usefulness is commonly not great, except as a brief introduction to more serious observation. The pupil should be taught to see accurately and in detail; and it is always well to lead him to make suggestions as to the meaning and uses of the features which he has seen.

In approaching the subject of nature-study, we must first ask why we desire to teach natural history subjects in the primary and secondary schools. There can be but two answers: we teach either for the sake of imparting the subject itself, or for the sake of the pupil. When we have the pupil chiefly in mind, we broaden his sympathies, multiply his points of contact with the world, quicken his imagination, and thereby deepen his life; a graded and systematic body of facts is of secondary importance. In other words, when the teacher thinks chiefly of his subject, he teaches a science; when he thinks chiefly of his pupil, he teaches nature-study. The child loves nature; but when he becomes a youth, and has passed the intermediate years in school, the nature-instinct is generally obscured and sometimes obliterated. The perfunctory teaching of science may be a responsible factor in this result. There seem to be four

chief requisites in nature-study teaching, if the pupil
is to catch inspiration from it:

1. The subject itself must interest the pupil. This
means that the instruction begin with the commonest
things, with those which are actually a part of the
pupil's life.

2. The pupil must feel that the work is his, and
that he is the investigator.

3. Little should be attempted at a time. One thought
or one suggestion may be enough for one day. The
suggestion that insects have six legs is sufficient for one
lesson. We obscure the importance of common things
by cramming the mind with facts. When the pupil is
taught to take systematic notes upon what the teacher
says, it is doubtful if the lesson is worth the while, as
nature-study. The pupil cannot be pushed into sympa-
thy with nature.

4. The less rigid the system of teaching and the
fewer the set tasks, the more spontaneous and, therefore, the better, is the result. A codified system of
examinations will choke the life out of nature-study.

In this nature-study, it would seem to be unwise to
rigidly grade the work, particularly as it is presented in a
text-book. The teacher can grade or adapt the matter,—he can fill out the frame-work,—as seems best for
his pupils and conditions. The work must be consecu-
tive, however, if it is to find a definite place in schools.
That is, some general plan or scheme must be laid out;

and in this direction it is hoped that this book of suggestions may be helpful. The first object of the book is to suggest methods, not to present facts. The liberal use of pictures in the book will suggest to the teacher the importance of having an abundance of illustrative material for the exercises, letting the pupils see the things themselves, as far as practicable, no matter how common or familiar they may be; and it is an advantage to have the pupils collect the specimens. The pupil's living contact with common things will strengthen the bond between the school and the home.

L. H. BAILEY.

HORTICULTURAL DEPARTMENT, CORNELL UNIVERSITY,
ITHACA, N. Y., December 13, 1897.

CONTENTS

I. TWIGS AND BUDS

(ix)

IV. PROPAGATION AND HABITS

V. COLLECTING

FIRST LESSONS WITH PLANTS

I. Twigs and Buds

I. THE BUD AND THE BRANCH

1. A twig cut from an apple tree in early spring is shown in Fig. 1. The most hasty observation shows that it has various parts or members. It seems to be divided at the point *f* into two parts. It is evident that the portion from *f* to *h* grew last year, and that the portion below *f* grew two years ago. The buds upon the two parts are very unlike and these differences challenge investigation.

2. In order to understand this seemingly lifeless twig, it will be necessary to see it as it looked late last summer (and this condition is shown in Fig. 2). The portion from *f* to *h*,— which has just completed its growth,—is seen to have only one leaf in a place. In every axil (or angle which the leaf makes when it joins the shoot) is a bud. The leaf starts first, and as the season advances the bud forms in its axil.

(1)

When the leaves have fallen, at the approach of winter, the buds remain, as seen in Fig. 1. Every bud on the last year's growth of a winter twig, therefore, marks the position occupied by a leaf when the shoot was growing.

3. The portion below *f*, in Fig. 2, shows a wholly different arrangement. The leaves are two or more together (*a a a a*), and there are buds without leaves (*b b b b*). A year ago this portion looked like the present shoot from *f* to *h*,—that is, the leaves were single, with a bud in the

FIG. 1.
An apple twig.

FIG. 2.
Same twig before leaves fell.

axil of each. It is now seen that some of these bud-like parts are longer than others, and that the longest ones are those which have leaves. It must be because of the leaves that they have increased in length. The body *c* has lost its leaves through some accident, and its growth has ceased. In other words, the parts at *a a a a* are like the shoot *f h,* except that they are shorter, and they are of the same age. One grows from the end or terminal bud of the main branch, and the others from the side or lateral buds. Parts or bodies which bear leaves are, therefore, branches.

4. The buds at *b b b b* have no leaves, and they remain the same size that they were a year ago. They are dormant. The only way for a mature bud to grow is by making leaves for itself, for a leaf will never stand below it again. The twig, therefore, has buds of two ages,—those at *b b b b* are two seasons old, and those on the tips of all the branches (*a a a a, h*), and in the axil of every leaf, are one season old. It is only the terminal buds which are not axillary. Buds are buds only so long as they remain dormant. When the bud begins to grow and to put forth leaves, it gives rise to a branch, which, in its turn, bears buds.

5. It will now be interesting to determine why

certain buds gave rise to branches and why others remained dormant. The strongest shoot or branch of the year is the terminal one (*f h*). The next in strength is the uppermost lateral one, and the weakest shoot is at the base of the twig. The dormant buds are on the under side (for the twig grew in a horizontal position). All this suggests that those buds grew which had the best chance,—the most sunlight and room. There were too many buds for the space, and in the struggle for existence those which had the best opportunities made the largest growths. This struggle for existence began a year ago, however, when the buds upon the shoot below *f* were forming in the axils of the leaves, for the buds near the tip of the shoot grew larger and stronger than those near its base. The growth of one year, therefore, is very largely determined by the conditions under which the buds were formed the previous year.

SUGGESTIONS.—At whatever time of year the pupil takes up the study of branches, he should look for three things: the ages of the various parts, the relative positions of the buds and leaves, the different sizes of similar, or comparable buds. If it is late in spring or early in summer he should watch the development of the buds in the axils, and he should determine (as inferred in 5) if the strength or size of the bud is in any way related to the size and vigor of the subtending (or supporting) leaf. Upon leafless twigs, the sizes of buds should also be noted, and the sizes of the former leaves may be inferred from the size of the leaf-scar (below the bud). The pupil should keep in mind the fact of the struggle for food and light, and its effects upon the developing buds.

II. THE LEAF-BUD AND THE FRUIT-BUD

6. Another apple branch is shown in Fig. 3. It seems to have no slender last year's growth, as Figs. 1 and 2 have at *f h*. It therefore needs special attention. It is first seen that the "ring" marking the termination of a year's growth is at *a*. There are dormant buds at *b b*. The twig above *a* must be more than one year old, however, because it bears short lateral branches at *e e*. If these branchlets are themselves a year old (as they appear to be), then the portion *f g* must be a similar branch, and the twig itself (*a f*) must be two years old. The ring marking the termination of the growth of year before last is therefore at *f*. In other words, a twig is generally a year older than its oldest branches.

7. The buds *c c* are larger than the dormant buds (*b b*). That is, they have grown; and if they have grown, they are really branches, and leaves were borne upon their little axes in the season just past. The branchlets *d d d* are larger (possibly because the accompanying leaves were more exposed to light) and *e e* and *g* are still larger. For some reason the growth of this

twig was checked last year, and all the
branches remained short. We find, in
other words, that there is no necessary
length to which a branch shall grow, but
that its length is dependent upon local or
seasonal conditions.

8. There are other and more important differences in this shoot. The buds
terminating the branches (*e e g*) are larger
and less pointed than the others are.
If they were to be watched as growth
begins in the spring, it would be seen
that they give rise to both flowers and
leaves, while the others give only leaves.
In other words, there are two kinds of
buds, fruit-buds and leaf-buds; and check-
ing the growth induces fruitfulness.

9. If the buds on the ends of the
branchlets *e e g* produce flowers, the twig
cannot increase in length; for an apple
is invariably borne on the end of a
branch, and therefore no terminal bud
can form there. If growth takes place
upon the twig next year, therefore, it
must arise from one of the lower or
leaf-buds. The buds upon the branchlets *d d d* will stand the best chance of
continuing the growth of the twig, for

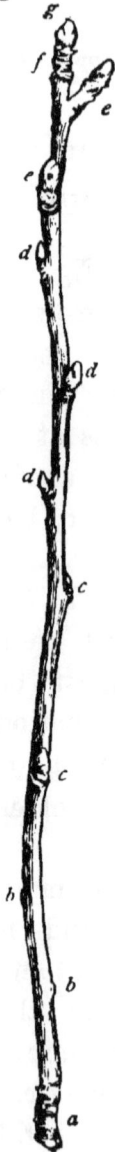

Fig. 3.
Formation of
fruit-buds.

they are largest and strongest, and are most exposed to sunlight. These failing, the opportunity will fall to one or both of *c c;* and these failing, the long-waiting dormant buds may find their chance to grow. In other words, there are more buds upon any twig than are needed, but there is, thereby, a provision against emergencies.

SUGGESTIONS.—The pupil should give himself some practice in determining or locating the rings marking the annual lengths of growth. A good way to do this is to choose some tree of known age (as a fruit tree or shade tree which has been planted but a few years), and endeavor to account for all the years' growths. He should also endeavor to find out how long the dormant buds may live upon any tree. He should attempt to determine if it is true that a moderate growth (so long as the tree remains healthy) tends to make the tree bear. Those persons who have access to vineyards should determine whether the most prolific canes are those of medium size and which do not run off to great lengths on the wires. Examine orchards for this purpose. Many pupils have heard that driving nails into trees tends to make them bear, and the result may have been attributed to some influence which the iron is assumed to exert upon the plant ; but if it is true that such practice induces fruitfulness, the pupil may be able to suggest an explanation of it. Let the pupil also determine whether dormant buds ever grow when the branch is injured above them.

It is evident, from the foregoing observations, that the twig in Fig. 3 cannot continue its growth in a straight or continuous line. Its terminal bud is to bear flowers, not to make a prolonged growth. The pupil should examine apple trees, or other plants, in which there are, occasionally or habitually, terminal blossom buds, and see how the plant increases in height. Perhaps he will notice that there is a tendency for the branches to fork. Let him see if the common red elder and the lilac make single or double terminal buds ; and if the latter, does that fact explain the forking and zigzag growth of these bushes ? What is the meaning of the forking growth of the sumac !

Some trees (as pines and spruces) continue to grow from the terminal bud, but most plants soon lose the terminal bud.

III.   THE STRUGGLE FOR EXISTÉNCE IN A TREE TOP

10. We have seen, in all the foregoing examples, that every twig bears more buds than can hope to find a chance to grow. Fig. 4 is an

FIG. 4.
The suppression of interior buds.

oak branch. It is seen that all the leaves are borne upon the very tips of the branches. That is, the interior of the space is poorly supplied with foliage. If the leaves are all borne at the ends

of the branches, then the branches must all arise
from the ends the following year, for we have
already found (2) that branches normally start only

Fig. 5.

The lengthening leaf-stalks on a horizontal shoot of Norway maple.

from leaf axils. The persisting branches, therefore,
may mark the general lengths of the previous annual growths.

11. Following the branches back we notice
that there are regular blank spaces and regular
points of branching. Every space between the
branches is a year's growth, but these spaces
still show the buds which failed to grow. Even
on the oldest part of the branch, the rough eleva-

tions where the buds were are still prominent; and
these scars may often be found on branches many
years old. The conclusion is that the method of
branching of a tree depends more upon the position of the buds with reference to light than it
does upon the position with reference to their
arrangement upon the twig.

12. Let the pupil lie under a dense shade tree
on a summer's day and look up into the dark
top. He will find that the interior of the top is
poorly supplied
with leaves, and
that the long
branches are leafy
at the ends. The
outside of the top
presents a wall of
foliage, often so
well thatched as
to shed the rain
like a roof, but
the inside is com-
paratively bare.
The tree may be
a maple. Fig. 5

Fig. 6.

Tip shoot of Norway maple.

is the tip of a side shoot. The lower leaves
have stretched out their stalks in eagerness for
the sunlight, for the newer leaves are constantly

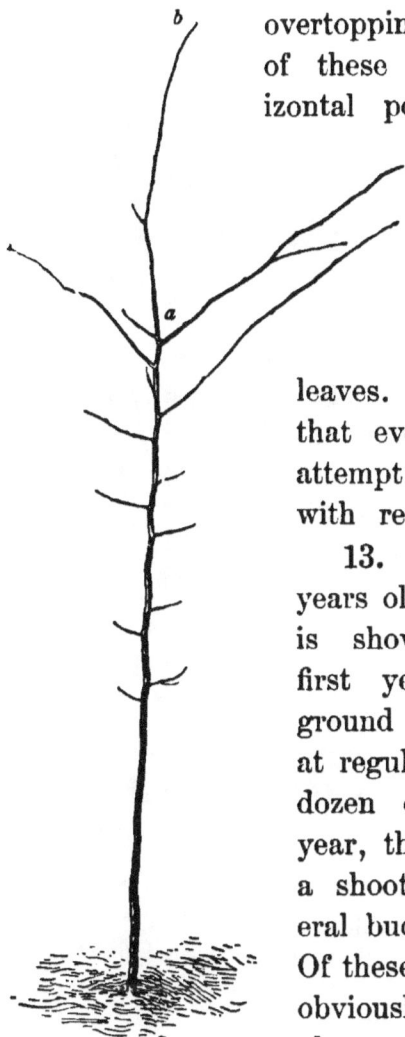

overtopping them; and the blades of these leaves stand in a horizontal position. Fig. 6 is a shoot from a topmost bough, where there is less struggle for light, and therefore shorter leaf-stalks and more various positions of leaves. It may be said, then, that even the leaves on a tree attempt to arrange themselves with reference to sunlight.

13. A black cherry tree two years old, taken from the woods, is shown in **Fig. 7.** The first year it grew from the ground to *a*; and it bore buds at regular intervals,—about two dozen of them. The second year, the terminal bud sent out a shoot to *b*, and thirteen lateral buds gave rise to branches. Of these thirteen lateral branches, obviously only three stand any chance of living in the dense shade of the forest. In fact, four or five of the lowest

Fig. 7.

The curious history of a wild cherry tree.

twigs were dead when the picture was made; showing that
the struggle for existence does
not always result from competition among fellows but may
arise from the crowding of
other plants.

14. These three strong
branches are less than four feet
from the ground, but other old
cherry trees standing near it
had no branches within fifteen
and twenty feet of the ground.
They no doubt branched low
down, as this one, but the
branches eventually died in the
struggle; and we therefore
have reason to conclude that
of all the branches on this
little tree, only the terminal
one, *b*, can long survive. The
trunk of a tree, then, is the
remainder in a long problem
of subtraction.

Fig. 8.

Indeterminate habit of the
sweet cherry.

SUGGESTIONS.—A young tree of the sweet garden cherry is
shown in Fig. 8, and one of the Morello or pie cherry in Fig. 9.
In the former, the terminal growths are strong, and the leader, or
central trunk, has persisted. The latter has long since lost its

leader, and the side growths are strong. Let the pupil now figure out how many buds have perished (or at least failed to make permanent branches) in each of these trees, if they are supposed to be seven years old. Any garden cherry tree will give him the

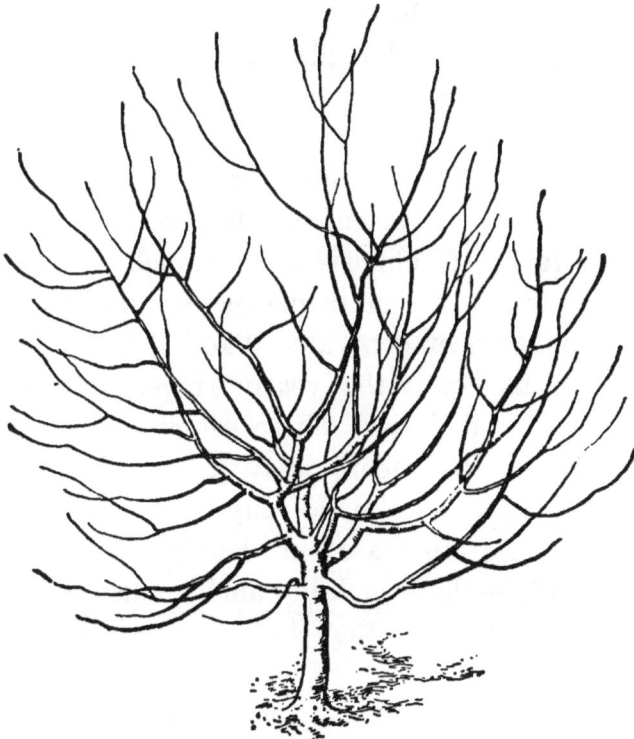

Fig. 9.

Determinate habit of the sour cherry.

probable number of buds to each annual growth. Even without the figures, it is evident that there are very many more failures than successes in any tree top. Let him also explain why the branches in Fig. 8 are in tiers.

IV. A BIT OF HISTORY

15. The apple shoot in Fig. 10 contains a volume of history. The illustration shows a single twig, but the branch is so long that it is broken several times in order to get it on the page. It arises at A, and continues, consecutively, at B, C, D, G, and F. A prominent feature of this shoot, —as, in fact, of almost any branch or plant,— is the presence of unlikenesses or dissimilarities. No two of the members are alike.

16. Let us count the yearly rings, and see how old the whole limb is. These rings are at 28, E, D, 12, 1,—five of them ; and as the shoot grew one year before it made any ring, and another year made no increase in length — as we shall see presently — the whole branch must be seven years old. That is, the limb presumably started in 1890.

16*a*. It is really impossible to tell whether the shoot started from the limb A in 1889 or 1890, without knowing the age of A; for the spur may have developed its blossom bud at the end in either the first or the second year of its life. That is, young fruit-spurs sometimes make a blossom bud the very year they start, but they oftener "stand still" the second year, and delay the formation of the blossom bud until that time.

We will begin, then, at A, and follow it out:

16*b*. 1890. Started as a spur from the main branch A, and grew to 1.

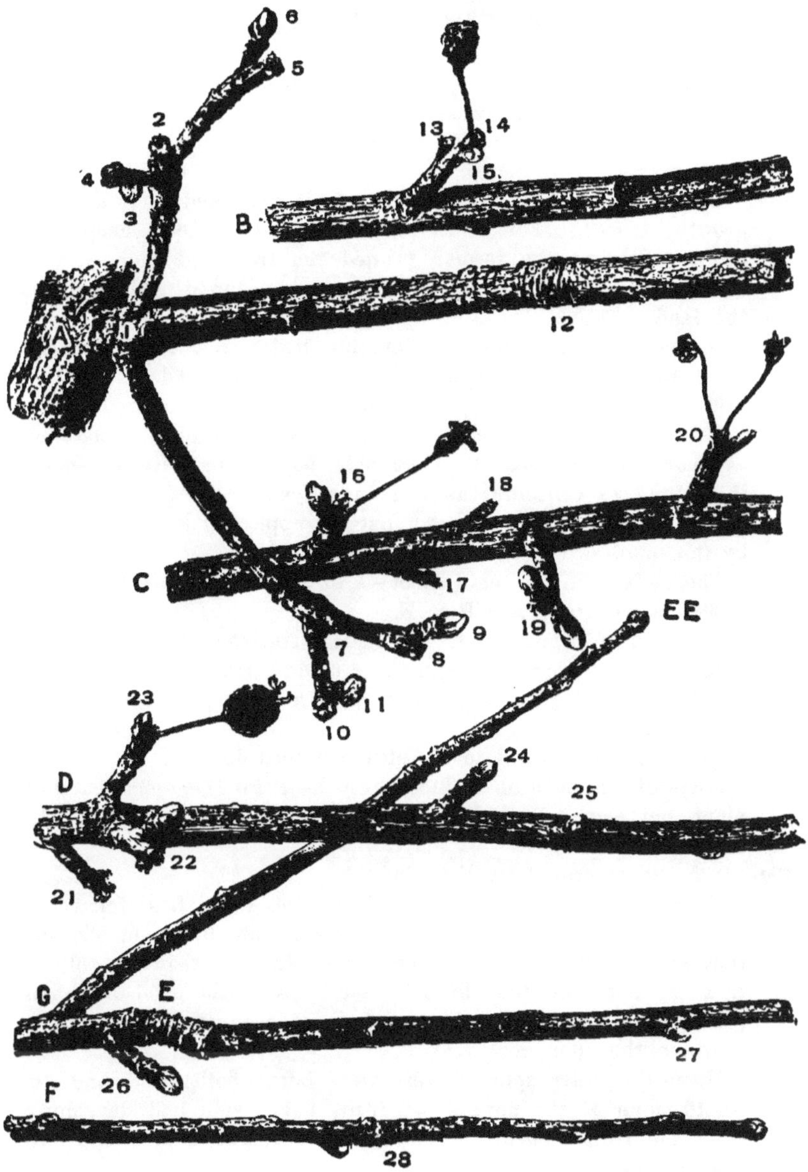

Fig. 10.

The eventful history of an apple twig.

16*c*. 1891. Apple borne at 1. This apple did not mature, however, as we can readily see by the smallness of the scar. In this year, two side buds developed to continue the spur the next year.

16*d*. 1892. Gave up its desire to be a fruit-spur, and made a strong growth, to 12. For some reason, it had a good chance to grow. Perhaps the farmer pruned the tree, and thereby gave the shoot an opportunity; or perhaps he plowed and fertilized the land.

In the meantime, one of the side buds grew to 3, and the other to 7, and each made a fruit-bud at its end.

16*e*. 1893. Shoot grew lustily,—on to D.

The fruit-bud at 3 bore an apple, which probably matured, as shown by the scar 2. Two side buds were formed beneath this apple to continue the spur next year.

The fruit-bud at 7 bloomed, but the apple fell early, as shown by the small scar. Two side buds were formed.

The buds upon the main shoot—1 to 12—all remained dormant

16*f*. 1894. Shoot grew from D to E.

Side bud of 2 grew to 4, and made a fruit-bud on its end; the other side bud grew to 5, and there made a fruit-bud.

Side bud of 7 grew to 10, and the other one to 8, each ending in a fruit-bud.

Buds on old shoot—1 to 12—still remained dormant.

Some of the buds on the 1893 growth—12 to D,—remained dormant, but some of them made spurs,—14, 16, 17, 18, 19, 20, 21, 22, 23.

16*g*. 1895. Shoot grew from E to 28.

Flowers were borne at 4 and 5, but at 4 the fruit fell early, for the five or six scars of the flowers can be seen, showing that no one of them developed more strongly than the other; that is, none of the flowers "set." A fairly good fruit was probably borne at 5. At the base of each, a bud started to continue the spur next year.

Upon the other spur, flowers were borne both at 8 and 10. At 10 none of the flowers set fruit, but a side bud developed. At 8 the fruit evidently partially matured, and a side bud was also developed.

The buds upon the old stem from 1 to 12 still remained dormant.

Some of the spurs on the 1893 growth — 12 to D — developed fruit-buds for bearing in 1896.

Some of the buds on the 1894 growth — D to E — remained dormant, but others developed into small fruit-spurs. One of these buds, near the top of the 1894 growth, threw out a long shoot, starting from G; and the bud at 26 also endeavored to make a long branch, but failed.

16*h*. 1896. Main shoot grew from 28 to the end.

The side bud below 4 (where the fruit was borne the year before) barely lived, not elongating, as seen above 3. This branch of the spur is becoming weak, and will never bear again. The side bud of 5, however, made a fairly good spur, and developed a fruit-bud at its end, as seen at 6.

The side bud of 10 grew somewhat, making the very short spur (11). This branchlet is also getting weak. The bud of 8, however, developed a strong spur at 9. Both 11 and 9 bear fruit-buds, but that on 11 is probably too weak ever to bear fruit again. In fact, the entire spurs, from 1 to 6 and 1 to 9, are too weak to be of much account for fruit-bearing.

This year several of the spurs along the 1893 growth — 12 to D — bore flowers. Flowers were borne from two buds on the first one (at 13 and 14), but none of the flowers "set." One of the little apples that died last June still clings to the spur, at 14. A side bud, 15, formed to continue the spur in 1897. Flowers were borne at 16, 20, 21 and 23, but no apples developed. Upon 16 and 20 the flowers died soon after they opened, as may be seen by the remains. Upon 23, one of the flowers set an apple, but the apple soon died. The spurs 17 and 18 are so weak that they never have made fruit-buds, and they are now nearly dead. The spurs 19 and 22 seem to have behaved differently. Like the others, they grew in 1894, and would have made terminal fruit-buds in 1895, and borne fruit in 1896; but the terminal buds were broken off in the fall or winter of 1894, so that two side buds developed in 1895, and each of these developed a fruit-bud at its end in 1896 in the spur 19, but only one of them developed such a bud in 22. Upon these spurs, therefore, the bearing year has been changed.

Upon the growth of 1894 — D to E — only three spurs have developed, Nos. 24, 25, 26. These started out in 1895, and two of them — 25 and 26 — have made large, thick buds, which are evidently fruit-buds. The shoot at G grew on to E E, and all the buds on its lower two-year-old portion remained dormant.

On the 1895 growth — from E to 28 — all the buds remained dormant except one, and this one — 27 — made only a very feeble attempt to grow into a spur.

The buds upon the 1892 growth — 1 to 12 — are still dormant and waiting for an opportunity to grow. Although these buds are five years old, they are still apparently viable, and would grow if they had the opportunity. Let the pupil determine how long these dormant buds may remain in apparently good condition on apple and other fruit trees.

17. What an eventful history this apple twig has had! And yet in all the seven years of its life, after having made fifteen efforts to bear fruit, it has not produced one good apple! The fault, therefore, does not lie in the shoot. It has done the best it could. The trouble has been that the farmer did not give the tree enough food to enable it to support the fruits, or he did not prune the tree so as to give the twig light and room, or he allowed apple-scab or some other disease to kill the young apples as they were forming. We may question, therefore, when trees fail to bear, whether it is not quite as often the fault of the farmer as the trees. Every orchard affords an interesting field for exploration, and even a youth may be able to dis-

cover facts which the fruit-grower himself may never have seen.

V. THE OPENING OF THE BUDS

18. We are curious to know how the buds of the apple twigs (Figs. 1, 2 and 3) open in the spring, and how the young growths start out. Let us look at the trees, and see. Fig. 12 is

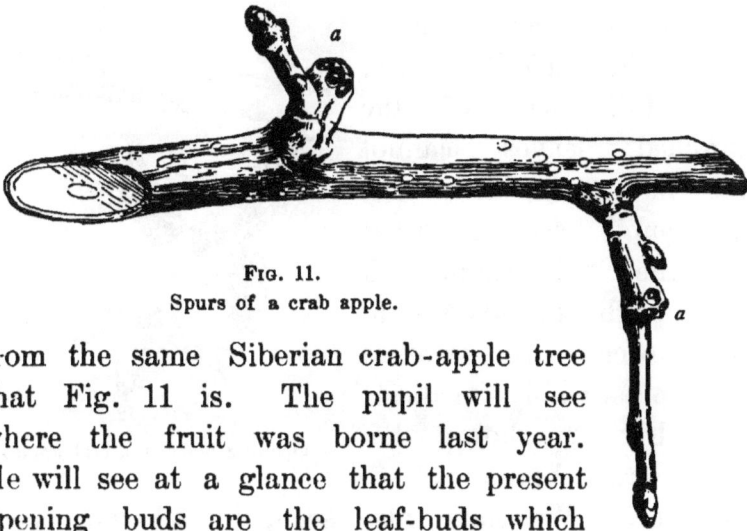

FIG. 11.
Spurs of a crab apple.

from the same Siberian crab-apple tree that Fig. 11 is. The pupil will see where the fruit was borne last year. He will see at a glance that the present opening buds are the leaf-buds which were formed on the side of the spur last year. The little dry scales which covered the buds in the winter have been pushed aside, and a new shoot is coming forth. The leaves are many. In a few days we shall be able to count them. Already nine of them are visible on the upper spur, and only eleven were borne all summer long on the annual growth in Fig. 2. The fact is that there are as many leaves packed away in the bud (as a rule) as there will be leaves on next year's shoot.

19. Another most curious fact about these opening buds is that the lowest leaves are smaller than the middle ones. The full size of the enfolded leaves cannot yet be made out. Let the pupil see to what size they will attain. It is enough to know that the lowest are smallest and presumably weakest; and Fig. 2 shows that they are borne closer

FIG. 12.

Opening buds of a crab-apple.

together. We have also seen, in all our specimens, that very few good buds are borne at the base of the annual growth (compare Figs. 1–3). We suspect, therefore, that not only the number of leaves, but the character of the forthcoming buds, is very largely determined beforehand.

20. These buds open with surprising quickness when spring comes (particularly at the north). The buds have been entirely inactive all winter, so far as we could see; and, moreover, they are just the same shape and size in the spring as they were when the leaves dropped in the fall. · We must conclude, then, that these leaves and an embryo shoot were packed away in the bud during the growing season of last year; and this is true.

21. The pupil can still further satisfy himself of the truth of this conclusion by taking into the house during the winter a twig from a tree or bush, and keeping it in water in a warm room. In a few weeks, it will produce leaves, and also flowers, if it bears flower-buds. This experiment also shows that the first leaves and flowers which come out on early spring-flowering trees and bushes are sustained by nourishment which is stored up in the branch or the bud, not by that taken in at the time by the roots.

22. Let the pupil examine a rapidly-growing shoot of any plant in spring or very early summer. He will not find the large buds which he sees in fall and winter. He concludes, therefore, that the plant does not need these large buds for purposes of growth. Plants must have a means of carrying the growing points over winter (or over the dry or inactive season, in the tropics); and in

order that time may be gained, the future branch
is packed away in miniature, ready to leap forth
upon the first awakening of spring.

23. If the leaves—and therefore the number of
buds on the shoot—are determined in the bud,
how does the shoot increase in length? If it grows
from its tip alone, the leaves would be left behind.
We know that this is not the case. Again, we
know that the joints or nodes (the places where
the leaves are borne) are really much closer together
in these opening shoots in Fig. 12 than they are
in the mature shoots in Figs. 1 and 2. In other

Fig. 13.

Opening of leaf-buds and flower-buds.

words, it is plain that the shoot increases in length
by elongating the internodes (or spaces between the
buds)

24. Another apple twig is shown in Fig. 13.
We are already familiar with the leaf-buds, but the
lowest bud is strange. It is a fruit-bud. The bud-
scales fall away, as before, but there comes forth

not only a cluster of leaves but a cluster of un-
opened flowers. We know that when an apple is
borne upon a spur, the spur ceases to grow in
that direction (p. 7); that
is, the apple fruit is
terminal. Then we know
that the shoot from this
bud is destined to re-
main short all summer,
and we infer that the
leaves upon this short
spur will exercise an im-
portant office in nourish-
ing the fruit.

25. We know that
apples are usually borne
singly, and yet the flow-
ers (as seen in **Fig. 13**)

FIG. 14.
Struggle for existence among the
apple flowers.

are in clusters. Two or three weeks after the flowers have gone, we examine the young apples, and
we see something like Fig. 14. One apple has
persisted and all the others have perished. There
is, then, struggle for existence even among flowers; and in apples, at least, we are to expect many
more flowers than fruits.

SUGGESTIONS.—The pupil should prove the conclusion in 14 ex-
perimentally. Let him lay off spaces at equal distances (say one-
quarter inch) on a young growing twig, and mark them with indelible
ink. If he visits the twig from day to day, and takes exact
measurements, he will make an interesting discovery.

II. LEAVES

VI. WHAT IS A LEAF?

26. Is there one leaf, or three, in the picture
of the dewberry (or blackberry), Fig. 15? We
have already found that branches
persist; that is, they do not
fall upon the approach of winter. Leaves commonly die and
fall. Here, therefore, is a means
of answering the question. Does
each of the three parts fall away
in the autumn and leave the
common stalk, *a*, upon the vine,
as a branch? Or
does the entire struc-
ture fall?

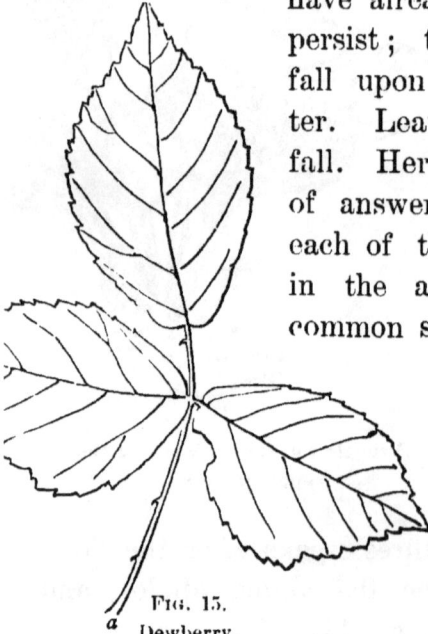

27. Buds are formed
in the axils of leaves,
as a rule. Where
are the axillary buds

FIG. 15.
Dewberry.

(24)

in the dewberry,—in the axils of each of the three parts, or in the axil of the stalk *a?* Again, leaves are borne at nodes; and the

FIG. 16.

Leaf of Lombardy poplar.

FIG. 17.

Spray of young apple leaves.

plant axis upon which they are borne either extends beyond them or gives evidence that it may do so.

28. What comprises the leaf in Fig. 16 (the

Lombardy poplar)? Is the leaf the expanded portion, or is it that portion plus the stalk? Let the pupil apply the above tests, and answer.

29. In the young apple foliage (Fig. 17), what comprises the leaf,—the expanded portion, the stalk, the two awl-like bodies at the base, or all of them together? How many leaves are there on the branch?

30. Point out the extremities of the leaf in the wheat (Fig. 18), or in any grass. Does it attach to the stem at 1 or at 6?

31. Designate the leaves in Figs. 19 and 20, and give the proofs.

VII. THE PARTS OF LEAVES

32. We are now ready to believe that a leaf may have two or three distinct parts,—the expanded portion or blade, the stalk or petiole, and appendages at the base, or stipules. We also know that it may have only the blade, as in the live oak leaves in Fig. 21; and it may have only the petiole or the stipules.

FIG. 18.
Leaf of wheat.

For example, observe that the bud-scales of the
common black currant gradually pass into leaves,
but the leaves are borne on the ends of scales;

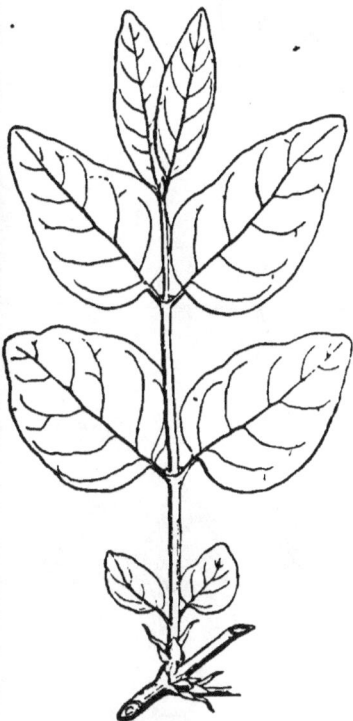

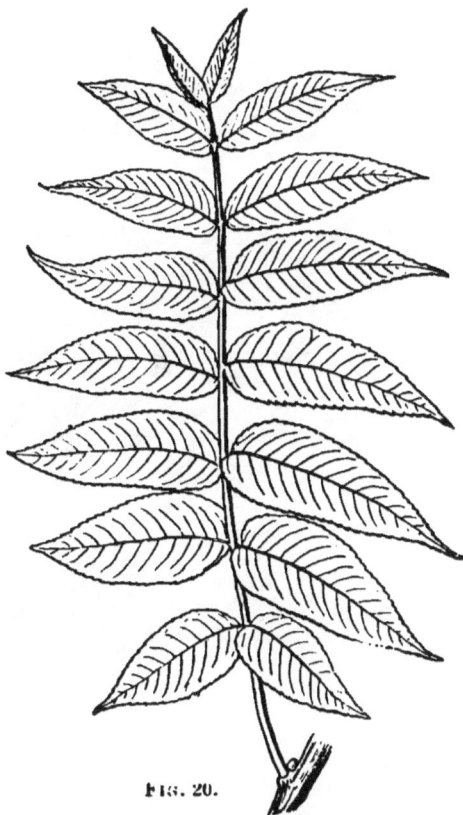

FIG. 19.

Honeysuckle.

FIG. 20.

Black walnut.

therefore, the scales must represent transformed
petioles, not transformed blades. The enlarged
and green bud-scales of the Norway maple are

petioles, affording an example of leaf-stalks which perform functions of leaf-blades.

FIG. 21.

Various leaves of a live oak.

FIG. 22.

Leaf of a willow.

33. The leaf of a willow is shown in Fig. 22. The stipules are so leaf-like as to indicate that they

must act the part of foliage (*i. e.*, perform the functions of green leaves). If, for any reason, the leaf-blades were to perish, it is conceivable that the stipules could maintain the plant. This actually

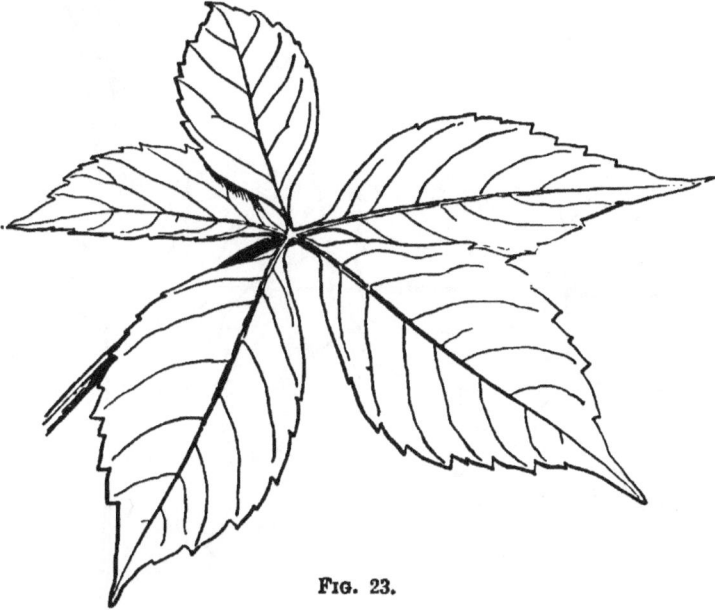

Fig. 23.

Virginia creeper.

occurs in some plants (as in some of the vetches), in which the entire foliage is made up of large stipules.

33a. A leaf which has no petiole is said to be sessile (*i. e.*, "sitting"), a term applied to any member which is destitute of a stalk or stem, as flowers, stamens, or fruits.

34. How shall we define the parts in the leaf of

the Virginia creeper (Fig. 23)?   The petiole is
plain;   but shall we say that there are five distinct
blades, or that the blade is divided into five parts?
Figs. 24 and 25 are leaves from one grape vine.

FIG. 24.

Grape leaf.

FIG. 25.

Deeply-lobed grape leaf.

Each plainly is one leaf.   The former has   three
well marked lobes, and the latter has these lobes
much more deeply cut.   In fact, there are strong
indications of five parts.   It is not difficult to imag-
ine the clefts extending to the mid-rib, as they

do in the Virginia creeper (which is a very closely related plant), and a compound leaf would be the result (that is, a leaf in which the blade is composed of at least two wholly separated portions).

35. Each part of the Virginia creeper leaf (and also of the dewberry leaf) is borne upon a distinct

Fig. 26.
Bean leaves.

stalklet of its own. These stalklets, then, are secondary petioles, or petiolules.

36. Bean leaves (Fig. 26) are seen to be compound, with both petiole and petiolules. Moreover, these petiolules are provided with little stipules, or stipels. Let the pupil now determine if there is a

joint at any place on the petiolules at which point the three parts may break off in the fall; and is the Virginia creeper like the bean in this respect?

37. The leaf of the Canada thistle (Fig. 27),— and of most other thistles,—is variously cut or jagged, but is nowhere completely separated, and is not, therefore, a compound leaf. We have seen, then, that there are various gradations between the simple leaf (that is, one in which the blade is one

FIG. 27.

Canada thistle.

more or less continuous piece, as in Figs. 16, 17, 18, 21, 22), and the compound leaf. In the true compound leaf the parts are generally articulated (or separated by joints), and are, therefore, usually provided with petiolules, although these are sometimes wanting. The different parts may fall independently of the entire leaf, or they may not.

38. Inasmuch as there seems to be a well

marked difference between the distinct divisions in the Virginia creeper and the ill-defined ones in grape and Canada thistle, we may give the two types different names. Or, the parts of a compound leaf are leaflets; the deep cut parts, like those in the thistle, are divisions or segments; the shallower parts (ordinarily not extending more than half way to the midrib) are lobes, as in Fig. 25.

SUGGESTIONS.—The pupil will now find himself applying the foregoing tests to all the leaves which he meets. Let him determine whether any plant bears both simple and compound leaves. He may be interested in examining the so-called Boston ivy or Japanese Virginia creeper which is much planted for covering houses; also, the horse-radish (examine the very earliest leaves in spring); also, one of the cultivated forsythias or yellow bells (the so called climbing one, Forsythia suspensa).

VIII. THE COMPOUND LEAF

39. The leaflets of the dewberry and Virginia creeper arise from a common point,—the top of the petiole. If the blade of the thistle (Fig. 27) were compound, the leaflets would evidently be distributed in two rows along a central axis. Compare Figs. 28 and 29. There are, then, two distinct types of compound leaves,—the digitate or palmate (in which the leaflets are at-

tached to a common point, like the bones of the hand to the wrist, and the petiole shows no tendency to continue beyond the point of their attachment); and the pinnate (in which the leaflets are arranged on the sides of an axis like the parts of a feather). The axis is prolonged in the pinnate leaves, and the part beyond the first leaflet is called a rachis.

40. If, now, the leaflets in Figs. 15, 23, 28 were grown together, what would be the method of attachment of the main veins or ribs in the resulting simple leaf? If the grape leaf (Figs. 24, 25) were to become compound, would it be palmate or pinnate? Would the oak (Fig. 21) have palmate or pinnate leaves? Why would the thistle leaf (Fig. 27) become pinnate rather than palmate? Let the pupil examine various kinds of leaves, and determine if simple leaves are either palmate-veined or pinnate-veined.

Fig. 28.

Leaf of poison ivy.

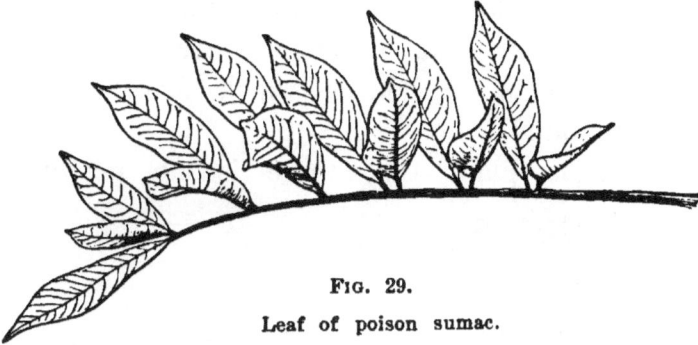

FIG. 29.

Leaf of poison sumac.

41. The Virginia creeper leaf has five leaflets, or is quinate (parts in fives). The dewberry and the poison ivy have three leaflets, or are ternate (part in threes). The jeffersonia (Fig. 30) has two leaflets, or is binate. Is this jeffersonia leaf essentially palmate, or essentially pinnate?

42. A leaf of the squirrelcorn (or dicentra) is shown, Fig. 31. It is evidently ternate and palmate; but each part is again divided into three, and each of these is again variously divided and cut. The leaf, therefore, is biternate (or twice ternate). The

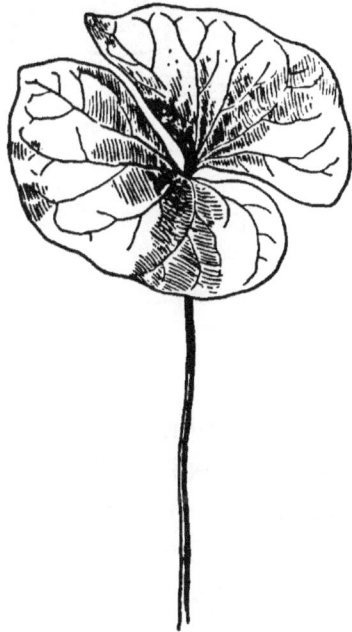

FIG. 30.

Binate leaf of jeffersonia.

entire leaf is said to be decompound (a term applied to all leaves in which the leaflets are compound; that is, to leaves which are more than
once compound).

43. It is plain that there is no positive or

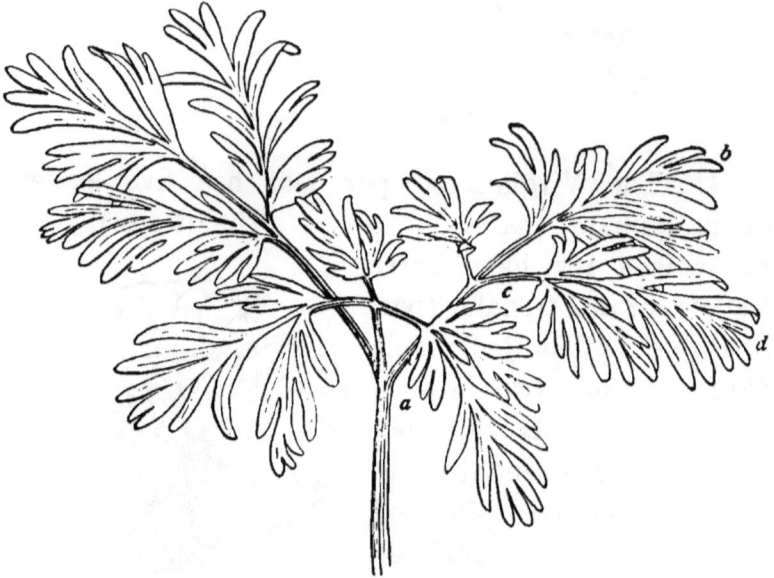

Fig. 31.
Leaf of squirrel-corn.

definite number of ultimate divisions in this
dicentra leaf. (Let the pupil examine the bleed-
ing-heart of the gardens, which is also a dicentra.) These ultimate parts are, therefore, not
leaflets, but segments or divisions. Is the leaflet the portion extending from *a* to *b*, or from

c to *d?* It is the latter; that is, it is customary, in speaking of decompound leaves, to use the term leaflet for the last part which is clearly and completely (and more or less uniformly) separated from its neighbors.

43*a*. The primary divisions in a palmately decompound leaf (as *a b*) are not given a distinct name in general botanical literature. The botanist would describe this dicentra leaf (Fig. 31) nearly as follows : Leaf ternately decompound (or sometimes written ternately compound, if the degree of compounding is afterwards specified), the main sections bearing palmately — or even pinnately — divided leaflets, the segments again deeply cut or divided.

44. The leaf in Fig. 32 (a gum arabic tree, a kind of acacia) is decompound, and is pinnate. Each of the numerous entire pieces or parts is called a leaflet, and the six primary parts are pinnæ. The leaf is pinnately bi-compound (or twice-compound). If each of the leaflets was again compound — which is not very rare in plants of this family — the leaf would be said to be tri-compound; the primary parts would still be called

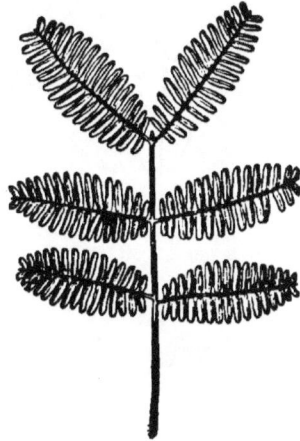

FIG. 32.

Twice-pinnate leaf of acacia.

pinnæ, the secondary parts pinnules, and the last complete divisions leaflets.

45. This acacia leaf has no terminal leaflets.

Compare the poison sumac (Fig. 29). That is, one is abruptly-pinnate, like the honey locust and the peanut (having no terminal leaflet), and the other is odd-pinnate. The latter is the more common form. Leaves are fairly constant

FIG. 33.

Spray of Currant tomato.

in these characters, but the pupil will be interested to find exceptions. Let him examine, among others, the leaves of black walnuts and butternuts.

46. Leaves of a tomato are shown in the

spray in Fig. 33. All the leaves have two kinds
of leaflets,—certain ones which may be taken as
the normal size, and other small ones interposed.
This kind of leaf is common in the tomato and
potato tribes. On account of the intermediate
leaflets, such a leaf is said to be interrupted.
This one, then, is interruptedly pinnate.

47. Another tomato leaf is shown in Fig. 34.

Fig. 34.
Leaf of Mikado tomato.

In this instance there are no interposed leaflets,
but the leaflets vary much in size and shape.
In other words, it is an example of an irregu-
larly compound leaf. The leaf looks as if it

might represent foliage of an indefinite form; or, in other words, that there is no absolute and typical form of tomato leaves. Let the pupil examine many tomato plants, and see if this is true.

48. The dahlia leaf is peculiar (Fig. 35). In this specimen there are five well-defined leaflets, C, O, M, M, A; but one of these, A, has given rise to a strong segment or division, and two others have divisions which are sufficiently distinct to be called leaflets. There are various grades of dividing or compounding, and the leaf may be said to be mixed. It is incompletely bi-compound.

49. From observations on leaves, we are soon impressed

Fig. 35.

Leaf of dahlia.

with the multitude of forms. We are also impressed with the fact that there may be great variety,—or elasticity,—of forms in the same kind of plant, showing that nature is really informal. Definitions, however, are formal; and it, therefore, follows that definitions should be compared with the objects, not the objects with the definitions.

The terminology (or naming) of compound leaves may be further explained, as follows:

49a. In making compounds to express the number of leaflets, the Latin for leaflet (*foliolum*) is used, not the word for leaf (*folium*); a trifoliate leaf is, therefore, an impossibility. It is like saying "three-leaved leaf." However, usage has sanctioned its employment, although it is etymologically improper. The better forms are—

Unifoliolate, a compound leaf of one leaflet;
Bifoliolate, of two leaflets;
Trifoliolate, of three leaflets;
Quadrifoliolate, of four leaflets;
Quinquefoliolate, of five leaflets;
Plurifoliolate, of several or many leaflets.

Any of these terms may be applied to either digitately or palmately compound leaves.

49b. The degree of compounding is often specified as follows:

Compound, once compound,
Bi-compound, twice compound, etc.;
De-compound, more than once compound, without specifying the degree.

Similarly, pinnately compound leaves may be designated as *bipinnate*, *tripinnate*, etc.; and palmately compound ones as *bipalmate*, *tripalmate*, etc. As a matter of fact, palmate leaves are rarely decompound if they have more than three primary divisions; so that it is customary to speak of palmately compound leaves as *ternate*, *biternate*, *triternate*, *multiternate*, etc.

SUGGESTIONS.—The venation of a leaf (or petal) is the arrangement and other features of the veins or ribs. Let the pupil collect abundantly of leaves (and indiscriminately, if he choose), and match the venation in them. Possibly he may find his pencil useful in recording and interpreting the differences. The drawings should be of use in the accustomed freehand exercises of the school; that is, the work with the pencil should be undertaken more in the interest of instruction in drawing than in the interest of nature-study.

IX. THE FORMS OF LEAVES

50. The forms of leaves (and of leaflets) interest us in two directions,— in respect to the relation which they bear to the welfare and history of the plant (or to adaptation to particular purposes of the plant), and in respect to their use in enabling us to recognize and describe plants. The former subject cannot be considered here. We shall, therefore, define the forms for purposes of description; but in doing this we must remember that there is every grade of intermediate form. Certain geometrical figures or arbitrary ideals are taken as the standards of comparison, and it must

Fig. 36.

Lanceolate leaves of
red pepper.

not be expected, therefore, that typical examples of the various forms are necessarily to be found in nature.

51. One of the first conceptions of forms of leaves which it is necessary to apprehend is that of the lanceolate (or lance-shaped) leaf. Lances were of

Fig. 37.

Ovate leaves of red pepper.

various shapes, but the botanical conception is a form four to six times longer than wide, and tapering at both ends, but the widest part is usually conceived to be below the middle. The leaves of the red pepper (Fig. 36) are examples.

52. Perhaps the next conception in importance

is that of the ovate leaf. This
is about twice as long as broad,
tapering from near the base to
a narrow or pointed apex. The
leaf at *a* in Fig. 37 (another
form of red pepper) is an example.

53. A third type form is
the oblong leaf. This is about
twice as long as broad, with
the sides nearly parallel from
top to bottom. Typical oblong
leaves are rare, but the form
is freely used in combination
with the lanceolate and ovate
types. Thus the chestnut leaf
(Fig. 38) is oblong-lanceolate.
The sumac leaflets (Fig. 29) are
ovate-oblong; so are the leaflets of Fig. 35. In these
combinations, the second word
is the one which is to be chiefly
emphasized; that is, an oblong-
ovate leaf is one which is more
ovate than oblong, whereas an ovate-
oblong leaf is one more oblong
than ovate. The narrower leaves
in Fig. 37 are lance-ovate (*i. e.*,
lanceolate-ovate).

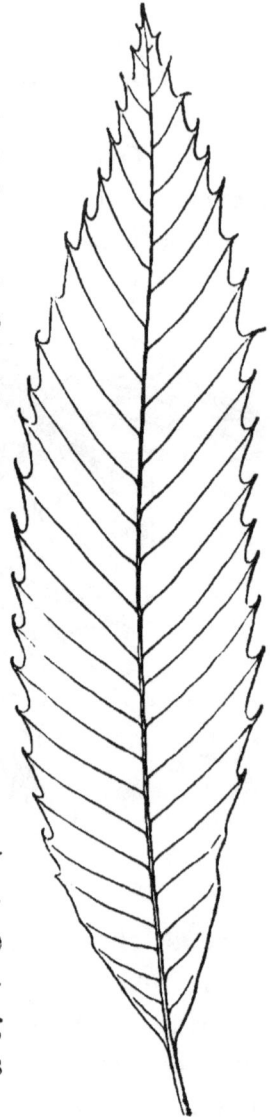

Fig. 38.

Oblong-lanceolate leaf of
chestnut.

54. Other type forms are the elliptical, which is like the oblong, except that it tapers equally both ways from the middle; spatulate, which is oblong with the lower end narrow; oval, which is broadly elliptical; orbicular, circular in outline; deltoid, or triangular; cuneate, or wedge-shaped;

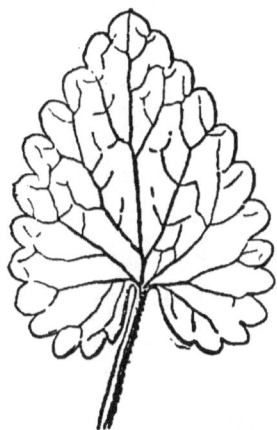

linear, or several times longer than broad, and the same width throughout; needle-shaped, as in pines and spruces. If any of the type forms are reversed, or inverted, the fact is expressed by the prefix *ob;* as oblanceolate, obovate. Combinations of these terms, together with the use of familiar adjectives (as short-ovate, long-lanceolate, round-obovate, etc.), express most of the common outlines of leaves.

FIG. 39.

Cordate-ovate crenate leaf of catnip.

55. Aside from the general outline, the form of the leaf is determined by the shape of its apex and base. The apex may be acute or ending in a sharp angle (Figs. 24, 25, 37); acuminate, ending in a long point (Figs. 26, 38); obtuse, or blunt (Fig. 19); truncate, or squared at the end; retuse, or indented (as the upper leaves in honeysuckle). The base may be cordate or heart-shaped

(as in Fig. 39, which is a cordate-ovate leaf);
reniform, or kidney-shaped; auriculate, or eared;
sagittate, or arrow-shaped; abrupt, or suddenly narrowed to the petiole (as in the broader leaves in
Fig. 37); gradually narrowed (as in Figs. 36, 38).
The cavity or recess in the base of a leaf, like
the grape (Figs. 24, 25), or moonseed is a sinus.

56. The features of the margins of leaves,
like their forms, are interesting because they are
intimately related to the origin or evolution of the
particular leaf (and, therefore, of the plant), and
also as a means of affording descriptive characters. The simple, straight margin is said to be
entire (Figs. 26, 29, 30, 36, 37). Departures from
this form are the serrate, or saw-toothed (Fig. 35);
dentate, or toothed (Figs. 23, 24, 25, 38, the last
being, perhaps, intermediate between serrate and
dentate); crenate, or scalloped (Fig. 39); repand,
or wavy, or undulate (Fig. 30 is obscurely so);
sinuate, which is a deep undulation; and then
follow the deep margins, as cut, jagged, lobed,
cleft, and the like. Leaves are said to be cleft
when the divisions extend deeper than the middle
of the blade, and lobed when they are not more
than half the depth of the blade.

56*a*. The diagrams of forms and margins of leaves given by
Linnæus are reproduced in exact form and size in Fig. 40: 1,
orbiculate; 2, sub-orbiculate (or subrotundate); 3, ovate; 4, oval,
or elliptical; 5, oblong; 6, lanceolate [narrower than present bot-

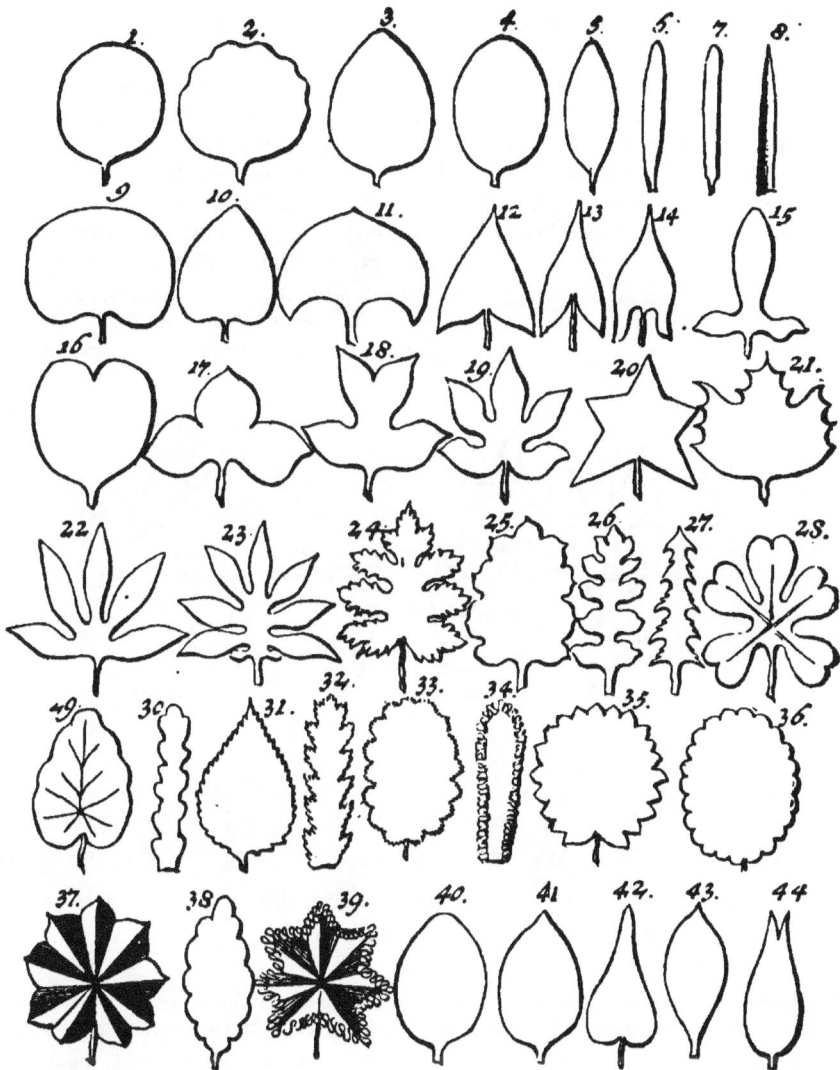

FIG. 40.

Linnæus' diagrams of leaves. 1751.

anists define lanceolate to be]; 7, linear; 8, subulate [awl-like]; 9, reniform; 10, cordate; 11, lunulate [or crescent-shaped]; 12, triangular; 13, sagittate; 14, cordate-sagittate; 15, hastate; 16, cleft ["fissum," now called obcordate]; 17, three-lobed, or trilobate; 18, premorse [irregularly notched at the end]; 19, lobed, or lobate; 20, five-angled; 21, erose [jagged or bitten]; 22, palmate; 23, pinnatifid; 24, laciniate; 25, sinuate; 26, dentate-sinuate; 27, retrorse-sinuate; 28, parted; 29, repand; 30, dentate; 31, serrate; 32, doubly-

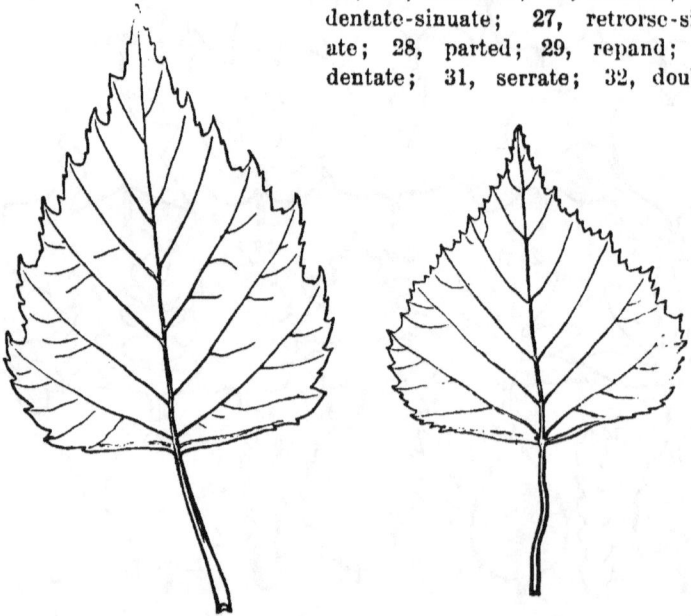

FIG. 41.

Variation in birch leaves from the same tree.

serrate; 33, doubly-crenate; 34, cartilaginous; 35, acutely-crenate; 36, obtusely-crenate; 37, plicate; 38, crenate; 39, crisped; 40, obtuse; 41, acute; 42, acuminate, ; 43, obtusely-acuminate; 44, emarginate acute.

SUGGESTIONS.—Let the pupil cut the form of any leaf in paper, and then endeavor to match it in other leaves. He will discover how difficult it is to describe a leaf with accuracy, and will also apprehend the greater truth that there are no two leaves alike.

III. Flowers

X. WHAT IS A FLOWER?

57. A flower of the hepatica, or liverwort, which springs from the mold with the first warmth of spring, is drawn in Fig. 42. The most hasty observation shows that it has several parts. Let us pull them away. We first find three green leaf-like members. Above these are several (seven in this case) pink or blue members. On the inside are about twenty hair-like bodies with pinkish enlargements on their ends, and each of these knobs seems to have two parts. Still inside, is a head of many greenish and pointed bodies. We know that the whole thing is a flower, but we are uncertain as to what

FIG. 42.

Flower of hepatica.

(49)

parts are most essential to it. A flower is obviously a more complex structure than a leaf.

58. A week or two later the flower has gone, and a structure like that in Fig. 43 has appeared in its place. We know that in the center of this structure are the seeds. We know, also, that the three green leaves will soon perish, as the other parts have perished, and only the little plants which spring from the seeds will bear testimony that there has been a flower. In other words, the purpose of a flower is to produce seeds, by which the plant is perpetuated.

59. If the above conclusion is true, it follows that the most essential or necessary parts of the flower are those which are directly concerned in the production of seeds. These parts, in the hepatica at least, are the very central organs. It is evident, therefore, that if we are properly to understand the flower, we must begin at the center, not at the outside.

FIG. 43.
After the flower is gone.

60. A flower of the common mustard is shown in Fig. 44. Secure a flower, and count the parts. The details (less half of the enveloping leaf-like parts) are displayed in Fig. 45. The central part, *o*, is to make the seed-pod. The minia-

ture seeds can be plainly distinguished if the part is held to the light. The mature seed-pod is shown in Fig. 46. This has grown to be so unlike the part *o*, that it is scarcely recognizable as the same member. It is necessary, therefore, for purposes of definition, to give the part, as it

FIG. 44.

Flower of mustard.

FIG. 45.

Details of mustard flower.

stands in the flower, a designative name. It is called the pistil.

61. This pistil is plainly of three parts,—the lowest and largest part, which bears the seeds, and which, therefore, we will call the ovary (or "egg-case"); the globular portion at the top, or the stigma (that is, a "mark" or "brand," in reference to its shape); the connecting portion, or style (in reference to its slender form).

62. Surrounding the pistil are four slender

bodies with enlargements at the top.   These are
shown at 1 1, 4 4, in Fig. 45.   The enlargements
are seen to have two parts, and
each part seems to have split along
its edge.   If the pupil were to
rub one of these enlargements upon
a bit of black paper, he would
probably discover a yellow dust.
These slender bodies are the stamens.   They are plainly of two
parts, the stalk, or filament, and
the enlargement, or anther; and
the anther contains the yellow pow-
der, or pollen, of which we have
spoken.

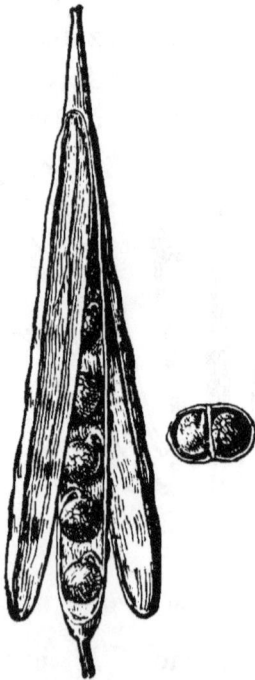

63.   There are two rows of
leaf-like parts surrounding the
pistil and stamens.   These are the
floral envelopes, or, collectively,
the perianth.   The inner row is
the colored or showy portion, or
corolla.   It has four parts, and

FIG. 46.

The seed-pod of the
mustard.

these we may call the petals.   It is suggestive
to note the similar forms of the petals and stamens.   Both have long stalks (technically called
claws in the petals) and a more or less expanded or enlarged portion at the top (the limb, in
the petals).

64. The outer row of the floral envelope or perianth comprises four smaller and greenish parts, which, individually, are known as sepals, and collectively as calyx. The calyx, corolla and stamens fall away and perish; and only the pistil matures into another member.

XI.   WHAT IS A FLOWER? CONCLUDED

65. If the pupil were to cut off the anthers before they open and discharge the pollen, and were then to cover the flower with a paper bag, or were to remove all other mustard flowers from the neighborhood, the pistil would soon die and fall. No seeds would be borne. It is, therefore, certain that the pollen is in some intimate way associated with the production of the seed.

66. If, however, having done this, the pupil were to bring pollen from another mustard flower and deposit it upon the stigma, he would find the pistil maturing and the seeds forming, as if he had not interfered with the flower. It is evident, therefore, that the office of the pollen is to cause production of seed by some action which it exerts after it is applied to the pistil. This action upon the forming seed is known as fertilization; and the transfer of the pollen to the stigma (whether

by the wind, insects, or by man) is pollination. There is a certain time when the stigma is receptive, or ready to receive pollen, and this condition comes when the pistil is full grown: the stigma then becomes viscid, or sticky, or much roughened, as if to hold the pollen. We now see that the stamens fall because they have performed their office; and the pistil persists that it may

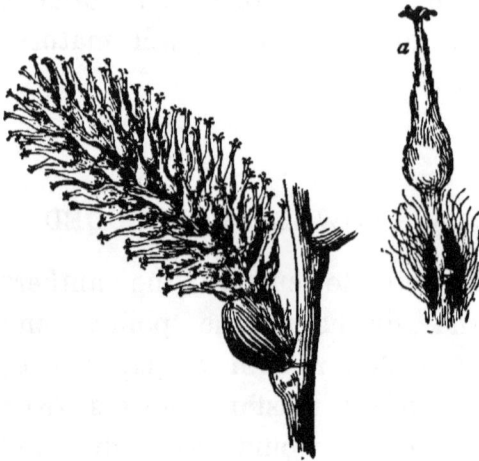

FIG. 47.
Pistillate flowers of willow.

mature the seeds. Since no seeds could be produced without the joint action of pistil and stamens, these members are known as the essential organs of the flower.

67. If the pupil were carefully to remove the petals and sepals, and were then to apply the pollen to the stigma, the pistil might mature and good seeds form. It is evident, then, that the floral envelopes do not hold the most vital relation to the office or purpose of the flower. They are not necessarily essential to it.

68. It so happens that in the greater number of plants the pistils and stamens in any flower mature at different times. That is, the pollen may all be discharged before the stigma is receptive, or the stigma may shrivel and die before the anthers open. In other words, there is frequently only a small chance of a flower fertilizing itself. There must be some means, then, of assuring the transfer of pollen. The commonest means are wind and insects. The flower does not need to attract the wind, but it must have some means of letting the insects know where it is. The showy petals are perhaps the sign-boards. At all events, insects may not visit some flowers when the petals are removed, although they are attracted by them when the petals are undisturbed.

68a. This non-concurrence in maturity of the essential organs is known as dichogamy.

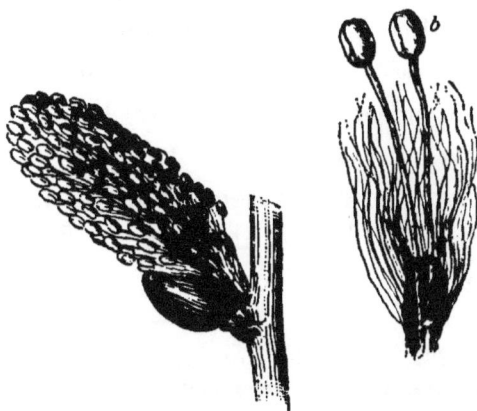

FIG. 48.

Staminate flowers of willow.

69. If the pollen may be carried from flower to flower, it is not essential that every flower

have stamens.   Figs. 47 and 48 are the soft
bodies which push out from the "pussy willows"
in spring.   They are really masses of flowers.
They are branches, since they are borne in the
axil of a bract or scale.   The cluster in Fig. 47
has members of a single kind, *a;* and these are
clearly pistils, since they bear an ovary and have
no pollen (no anthers).   The cluster in Fig. 48
also has members of a single kind, *b*, but they
are unlike the members of Fig. 47.   They are
stamens, as may be determined by the pollen and
the filaments, and the absence of ovary.   In both
cases, the parts have no envelopes, but are borne
in the axil of a hairy or woolly scale ; and it
is this silky wool which gives the name of "pussy
willow" to the plant.   Such flowers are said to be
imperfect, because they have only stamens or pistils, in distinction to the perfect flowers, which
have both stamens and pistils.

70. What, then, is a flower?   It is essentially
only a pistil or a stamen.

70a.   Since the flower may have two kinds of envelopes—and two
kinds of essential organs—it is commonly said that the complete flower
is one which has all of these parts, and an incomplete flower is one
in which one or more of the series is missing; but this is only a
method of stating one's habit of thinking about a flower, and it may
lead the beginner to think that there is some necessary or typical
plan of flower from which most flowers are deviations.   It would be
better to drop the terms complete and incomplete, and to say that

flowers which have all the four parts are quadriserial; those lacking
only the calyx are triserial; those lacking the floral envelopes are bi-
serial; and those which contain only the pistil are uniserial.

70*b*. It is customary, however, to speak of flowers which lack the
calyx as asepalous; and of those which lack the corolla as apetalous.
When flowers lack both the calyx and corolla (as the willows), they
are said to be naked, or achlamydeous.

70*c*. Flowers which contain pistils and no stamens are said to be
pistillate, or fertile. Those which have stamens and no pistils are
staminate, or sterile. In common language they are sometimes said to
be female and male, respectively, but the former terms are better
when speaking of the parts as facts (or as members), without reference to sexuality. When pistillate or staminate flowers are spoken of
without designating which they are, they are properly said to be di-
clinous; which is essentially the same as to say that they are imperfect, as this term is generally used. They are sometimes said, also,
to be unisexual, in distinction to bisexual or hermaphrodite flowers
(which have both stamens and pistils).

70*d*. When speaking of the staminate portion alone, it is customary to call it the andrœcium; and to call the pistillate portion the
gynœcium.

SUGGESTION.—The pupil should now have practice in distinguishing
the members or parts of flowers, and in interpreting the unusual or
disguised parts.

XII.   THE PARTS OF THE PISTIL

71. The pistils of hepatica, mustard, tulip
(Fig. 49), and willows are composed of a
single straight column. The mustard and willow have a distinct style, but the hepatica and
tulip differ in having none. That is, the stigma
is often sessile on the ovary, from which we

conclude that while the ovary and stigma are essential to a pistil, the style is not.

72. In all the flowers which we have so far examined the style is single; that is, there is only one straight style on each ovary. In the apple, however (Fig. 50), the styles are five, while the ovary is but one. The pupil should now examine any flowers which he meets, with respect to the absence or presence of styles and to their number; and he will find variations from none whatever to several, or even many, to a single ovary.

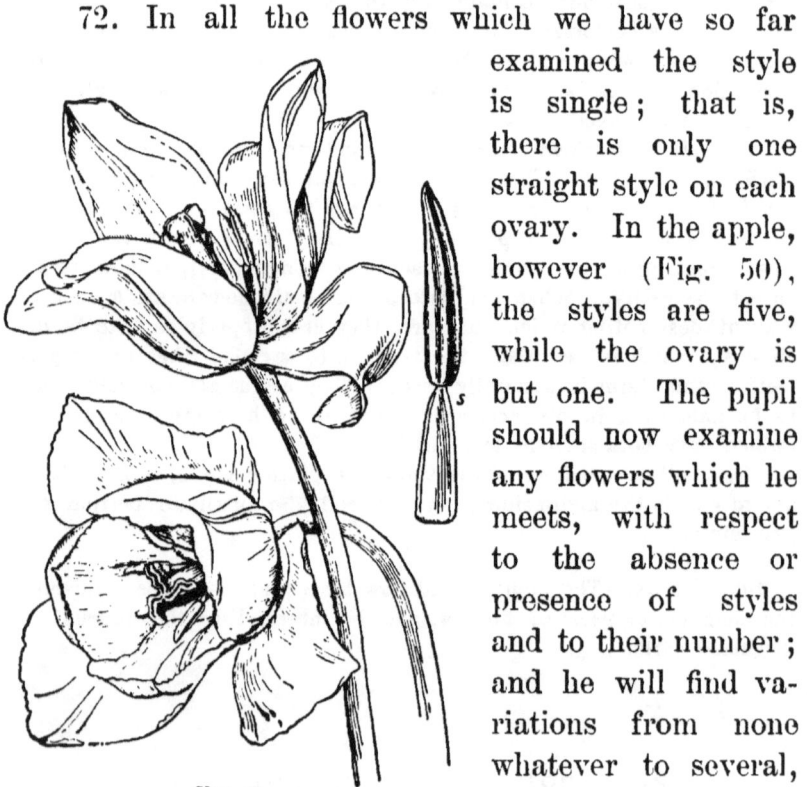

Fig. 49.

Flowers of tulip.

73. In the hepatica, mustard and apple, the stigma is one for each style; in the tulip there are three stigmas (or at least three parts to one stigma); in the willow there are two stigmas, and each is again two-parted, and in the catnip (Fig.

51) there are two. These stigmas differ not only
in number, but in size and shape. We conclude,

FIG. 50.
Flowers of the apple.

therefore, that stigmas
acteristic forms, as
74. If we examine
tard (*o*, Fig. 45), we
comprises two distinct
seeds in each. The
enlarged in the ma-
show their character
are borne inside the
true of most plants),
compartments which we

have peculiar and char-
styles do.
the pistil of the mus-
observe that the ovary
compartments, with
parts are sufficiently
ture pod (Fig. 46) to
well. The seeds, then,
ovary (and this is
in distinct cavities or
may call locules.

FIG. 51.
Pistil of
catnip.

74*a*. The compartments of
cells, but this is a common-
has prior use in general lit-
in botanical writings, it should,
ignate the ultimate structural

ovaries are commonly called
language word, and therefore
erature. If it is used at all
perhaps, be restricted to des-
elements or units of the plant,

as employed by anatomists and physiologists. Locule is an-glicized from loculus, diminutive of Latin *locus,* "a place."

FIG. 52.

Ripened pistil of hepatica.

FIG. 53.

Cross-section of ovary of tulip.

75. In the hepatica Figs. 42, 43), there are several dis-tinct pistils in a head. Each one contains but a single locule, and ripens but a single seed (Fig. 52). The pistil of the tulip, however (Fig. 53), has three locules, corresponding to as many sides or angles. Pistils contain different num-bers of locules, according to the kind of plant of which they are a part.

75*a.* Pistils with one locule are unilocular or 1-loculed; those with two are bilocular or 2-loculed; those with three, trilocular or 3-loculed; those with four, quadrilocular or 4-loculed; those with five, quinquelocular, or 5-loculed; those with several or many, mul-tilocular, or ∞-loculed.

76. The ovary is not only variously divided into compartments, but the ovules (or bodies which mature into seeds) are attached to different parts of the locule. In the mustard they are attached to the central partition of the ovary, in the tulip to the interior walls of the locules, in the corn-cockle (Fig. 54) to a columnar central portion, and in the plum (Fig. 55, *o*) to the outward side of the locule. In general, there is a more or less distinct elevation or thickening of tissue at the

place where the ovules are attached. This is emphatically shown in the fruit of the May-apple or mandrake (shown in cross-section in Fig. 56). This point of attachment is known as the placenta (plural, placentæ).

76a. The placenta is defined with reference to its position. It is evident that there are two general types of placentæ,—those which are borne upon the outward walls of the ovary, and are called parietal, and those that are borne in the center, and are called axile. Of the axile placentæ, there are two kinds, those which are attached to the partitions or dissepiments of the ovary (as in the tulip, Fig. 53), and those which are borne upon a separate central column, and are, therefore, called free axile placentæ (as in the cockle, Fig. 54, and in all the pink tribe, as the pinks, carnations, chickweeds, catchflies; and also in the primroses).

77. It is now seen that the pistil is not

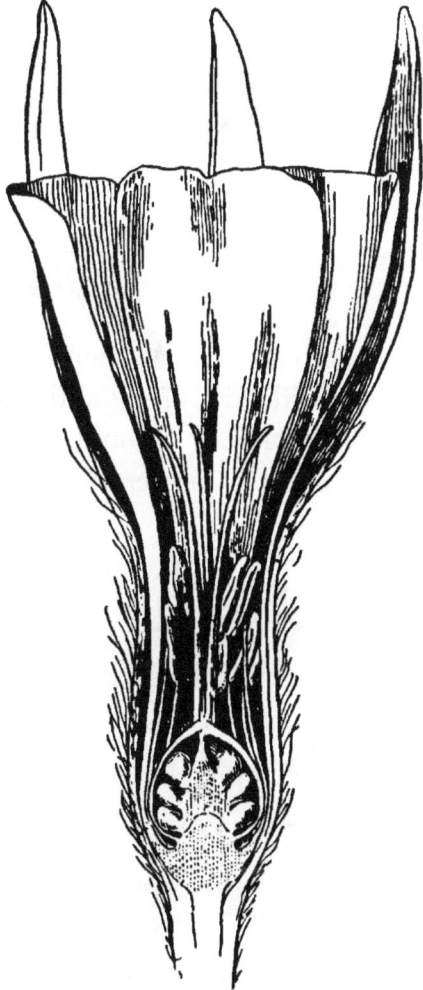

FIG. 54.

Free axile placenta of corn-cockle.

always the simple structure which it looks to be
from the outside. That is, it may be either simple

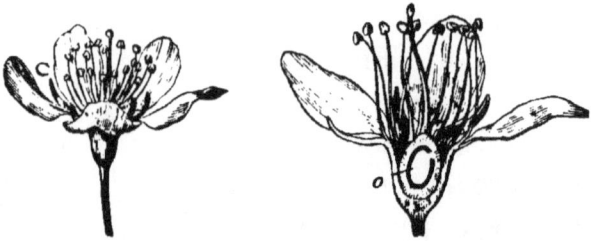

Fɪɢ. 55.

Flowers of plum.

or compound. A compound pistil is one which
bears evidence of containing two or more united
parts or units. The common test of a compound
pistil is the presence of more than one locule,
but this is not always des-
ignative, for in some cases
false partitions grow out
from the walls into the cavity of the ovary. The presence of more than one
style to a single ovary also
indicates a compound pistil;
and, more especially, the
occurrence of more than
one placenta. The separable

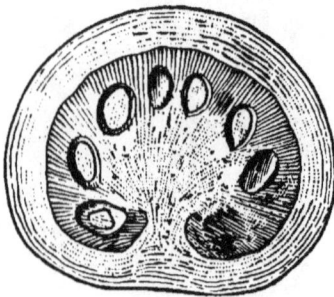

Fɪɢ. 56.

Large parietal placenta of
may-apple.

units or parts in a compound pistil are known as
carpels. The theory of a compound pistil is that

it is made up of the union of two or more simple
pistils.

77a. Thus the hepatica has one carpel, the tulip has three, the
mustard has two, the catnip has two 2-lobed carpels, the apple has
five, and even the unilocular cockle (Fig. 54) is thought to be
5-carpelled because of the five styles (two being cut away in the
figure) and of certain peculiarities in related plants;—that is, there
is evidence that some plants which were once 5-loculed are now
1-loculed because of the loss of partitions ; and sometimes this
elision can be traced in the different ovaries of a single plant.

77b.. A flower, therefore, may contain one simple pistil, several
simple pistils, or one compound pistil; and there are instances in
which it contains more than one compound pistil.

SUGGESTIONS.—When taking up any unfamiliar flower, look first
for the pistil. The ovary is the best distinguishing mark, for the
pistil is often much disguised. Determine what relation exists
between the numbers of stigmas, styles, or locules in any pistil.
Also observe the number of ovules, and the placentæ.

XIII. THE STAMENS

78. The most striking feature of the stamens
in the flowers which we have seen
is the great difference in length and
shape. Most of the stamens are slender, and have prominent stalks or
filaments ; but the anthers of the
currant (Fig. 57) are nearly sessile,
and in some flowers they are completely sessile. It is, therefore,

FIG. 57.

Flower of garden
currant.

apparent that a filament is not essential to a stamen
any more than a petiole is essential to a leaf.

79. All these an-
thers appear (so far
as we can see) to con-
tain more than one
cavity. Most of them
apparently have two
compartments; and
this is the general
rule. It is easy to
ascertain that these
compartments (which

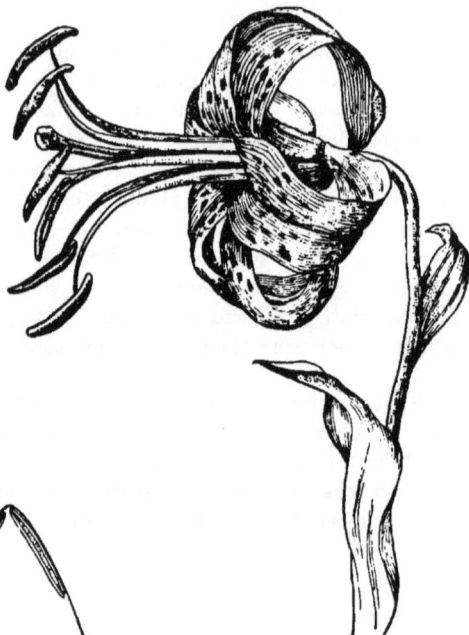

Fig. 59.

Flower of wild lily.

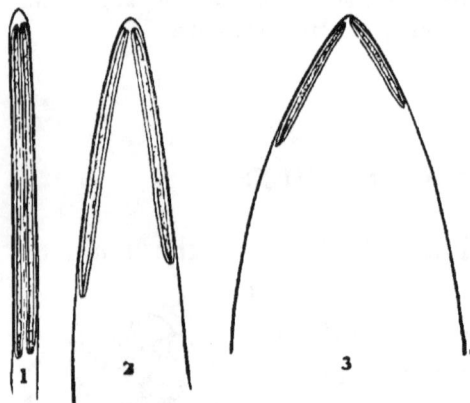

Fig. 58.

Stamens of water lily.

we shall call lo-
cules) contain the
pollen (62).

79a. It is the custom
to call the anther com-
partment a cell, but this
word should be other-
wise employed (74a). As no confusion has arisen from the application of the word cell to both pistils and stamens, none may
be anticipated from a like use of locule. It has been suggested

to use locellus (diminutive of loculus) for the anther compartment, but it seems to be unnecessary to introduce another word, and, moreover, locellus has no accepted anglicized form (although it might be shortened to locel).

80. The anther of the tulip and willow is attached by the base to the very top of the filament, but that of the water-lily (Fig. 58) seems to be joined to the filament in its entire length. The mustard and the lily (Fig. 59) show still a third method, the anther being poised by attachment to its back, and standing cross-wise the filament. These three methods, with numerous intergradations, will impress the pupil, if

Fig. 60.

Pores in azalea stamen.

he were to examine numbers of flowers, as being the types of the ways in which the anther is borne upon the filament.

Fig. 61.

Sensitive stamens of barberry, showing a single flower, and the dehiscence of the anthers at *a* and *d*.

80*a.* These modes may be called, respectively, the innate (attached at base), adnate (attached throughout its length), and versatile (attached near the middle, or at least at some distance from the ends).

81. The exposure of the anthers in the mustard and the lily is in opposite directions. The anthers of the mustard

look inward (towards the pistil), or are said to be introrse; those of the lily look outwards, or are extrorse. The pupil should determine if innate and adnate anthers differ in this regard, also.

FIG. 62.   FIG. 63.   FIG. 64.

Flower of scarlet sage.   Parts of carnation.   Stamens in hepatica.

82. The anthers of the mustard and the tulip seem to open along the side of each locule. The azalea, however (Fig. 60), opens by a hole or pore in the tip of the locule. Heaths and huckleberries open in the same way. We should examine

the barberry (Fig. 61), in which the anther opens
by means of a lid.

82*a*. The barberry flowers are honey-sweet, and attract the bees;
and the plant seems to make the most of its opportunity. When the
flowers are just expanded, and the sun is warm, touch the filaments
upon their inner side with a pin or point of a pencil. See what
happens. Observe, also, the curious way in which the anthers open.
The pupil will now be interested in the anthers of other plants of
this family, such as may-apple, jeffersonia, and blue cohosh.

82*b*. The opening of any closed organ is known as its dehiscence. We have found, then, that the dehiscence of the anther
locules is various, and that it follows at least three types or methods.

83. We have seen that there are commonly
two locules, and in the water-lily (Fig. 58) they
are separated by the width of the filament. A
flower of the scarlet sage of gardens and green-
houses is laid open in Fig. 62. The anthers are
at 1 and 2; but a closer examination of the
anther shows that it has but a single locule, and
as other mints have two, we are suspicious that
the other compartment has been lost. The truth
is that in some kinds of sage (as the common
garden sage) the two locules are separated by a
stalk or bar, which runs crosswise the top of the
filament. This bar, separating the two locules of
an anther, is called a connective. In the flower
before us, the other locule has apparently vanished in the process of time, and the places where
we should expect to find it are at 3 and 4, on

the other end of the connective. We have, then,
still a fourth kind of anther-bearing, but it is
clearly a special case of versatile arrangement; that
is, it is not a general type or mode.

SUGGESTIONS.—The presence of pollen is the one infallible proof
of stamens. The pollen is commonly in the form of yellow grains,
and is easily recognized even by the naked eye. In identifying
stamens, note first the form and dehiscence of the anther, then the
position of the stamen with reference to other parts of the flower.
Find the stamens of the apple, rose, strawberry, carnation, lily,
crocus, lilac, honeysuckle, verbena, orange, fuchsia, geranium.

XIV. THE DANDELION

84. The first warmth of spring brought the
dandelions out of the banks and knolls. They
were the first proofs that winter was really going,
and we began to listen for the blackbirds and
swallows. We loved the bright flowers, for they
were so many reflections of the warming sun.
They soon became more familiar, and invaded the
yards. Then they overran the lawns, and we
began to despise them. We hated them because
we had made up our minds not to have them, not
because they were unlovable. In spite of every
effort, we could not get rid of them. Then if we
must have them, we decided to love them. Where
once were weeds are now golden coins scattered

FIG. 65.
Dandelion.

FIG. 66.
Floret of dandelion.

in the sun, and bees revelling in color; and we
are happy!

85. A dandelion is shown in Fig. 65. It is
a strange flower, as measured by those which
we have already studied. It appears to have a

calyx in two parts or series, and a great number
of petals. If we look for the pistils and stamens,
however, we find that the supposed simple flower is
really complex. Let us pull the flower apart and
search for the ovary or seed. We find numerous
objects like that in Fig. 66. The young seed is
evidently at *c*. There are two styles at *d*, and a
ring of five anthers at *b*. The dandelion, therefore,
must be composed of very many small and perfect
flowers.

86. Looking for the floral envelopes, we find a
tube, and a long strap-like part running off to *c*.
This must be corolla, for the calyx is represented by
a ring of soft bristles, *a*. We have, then, a head
made up of quadriserial flowers, or florets, as the
individual flowers may be called. The entire head
is reinforced by an involucre, in much the method
in which the dogwood is subtended by four petal-
like bracts and the calla spadix by a corolla-like
spathe.

87. One cloudy morning the dandelions had
vanished. A search in the grass revealed num-
bers of buds, but no blossoms. Then an hour
or two of sunshine brought them out, and we
learned that flowers often behave differently at
different times of the day and in various kinds
of weather.

88. In spite of the most persistent work with

the lawn mower, the dandelions went to seed profusely.
At first, we cut off many of
the flower-heads, but a
the season advanced
they seemed to escape
us. They bent their
stems upon the ground
and raised their heads
as high as possible and
yet not fall victims
to the machine; and
presently they shot
up their long soft
stems and scattered
their tiny balloons to
the wind, and when the
lawn-mower passed,
they were either ripe
or too high to be
caught by the machine.

89. This seed has
behaved strangely in
the meantime. The
fringe of pappus (as
the bristle-like calyx
is called) is raised
above the seed by

Fig. 67.

The dandelion.

FIG. 68.

Variation in dandelion leaves. All drawn natural size and then
reduced one-half.

a short, narrow neck (*c*, Fig. 66), when the plant is in flower; but at seed-time this neck has grown an inch long (Fig. 67), the anthers, styles and corolla have perished, the pappus has grown into a spreading parachute, and the ovary has elongated into a hard, seed-like body. Each one of us has blown the tiny balloons from the white receptacle, and has watched them float away to settle point downwards in the cool grass; but perhaps we had not always associated these balloon voyages with the planting of the dandelion.

90. The dandelion, then, has many curious habits. It belongs to the great class of compositous (or compound) flowers, which, with various forms, comprises about one-tenth of all the flowering plants of the earth. The structure of these plants is so peculiar that a few technical terms must be used to describe them. The entire "flower" is really a head, composed of florets, and surrounded by an involucre. These florets are borne upon a so-called receptacle. The plume-like down upon the seeds is the pappus. The anthers are said to be syngenesious ("in a ring"), because united in a tube about the style; and this structure is the most characteristic feature of compositous flowers,—more designative of them, in fact, than the involucrate head, for in some other kinds of plants the flowers are in such heads, and in

some compositous flowers the florets are reduced
to two or three, or even to one!

SUGGESTIONS.—Is the bud at the right in Fig. 67 a flower
closed up, or one which has not yet opened ? Are the stems of
the dandelions which bloom first in the spring shorter than those
which bloom later ? Do the flowers close at night and in dull weather ?
How long a period of sunshine is necessary to open the flowers ?
Does a flower open more than once ? Does the head (or involucre)
ever close up after it has gone to seed ? What time is required
for the flower stem to straighten up and to reach its full height ?
How many rows of bracts or scales are in the involucre ? Do the
positions of these bracts change from flowering-time to seeding-
time ? How far may a dandelion seed travel in the wind ? Do
dandelion plants vary much in size and shape of leaves (compare
Fig. 68)? Is the variation associated with vigor of plant, richness
and moisture of soil, or other conditions ? At what seasons are
dandelions most abundant ? Do they ever bloom in fall or winter ?
How long does a dandelion plant live ? Upon what kind of soil
does it thrive best ?

XV. CROSS-FERTILIZATION

91. We have found (58) that the purpose of
the flower is to produce seeds ; these seeds cannot be formed without the aid of pollen; comparatively few flowers are perfect and also synanthous
(or simultaneous) in the maturation of pistils and
stamens, and very many flowers are imperfect.
It would seem to follow, therefore, that cross-fer-
tilization is the rule, and we infer that it must
result in some decided benefit.

92. The simplest means by which cross-fertili-
zation is enforced is by dichogamy, or the different

times of maturing of the organs of the same flower (68*a*). Certain simple movements or habits of the pistils or stamens are often associated with dichogamy. Fig. 69 is a flower of one of the wild phloxes. The stigmas are seen to be three, but these are closed until the stigmatic surfaces are receptive, which commonly occurs after the pollen is discharged. A similar behavior may be detected in campanulas or blue-bells. In the young flowers the style is merely club-shaped; in the oldest flowers, the style has opened to three branches, but the anthers are shrivelled. Inasmuch as the period of blooming of any

FIG. 69.

Dichogamous flower of phlox.

plant usually extends over several days at least, the dichogamous flower is likely to receive pollen from various flowers which are borne either upon the same or another plant.

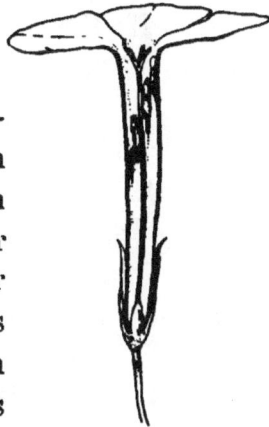

92*a.*. Pistils of dichogamous flowers may accidentally receive pollen from the same flower; but Darwin and others have found that pollen is often impotent, or sterile, upon the associated stigmas. That is, if pollen from the same and from another flower were to fall upon a stigma, the foreign pollen is the more likely to be fecund. Foreign pollen is commonly prepotent. If, however, no pollen is received from another flower, the stigma may accept the pollen from the associated anthers.

93. It is evident that if self-fertilization is so
often excluded, the plant must frequently depend
upon extraneous agents for the transfer of pollen
and the perpetuation of its kind.

94. If the pupil were to shake the staminate
catkins of the hazel, birch or walnut when they
are mature, he would be surprised at the showers of
pollen which are discharged; and if he should
watch the destination of this pollen he would
probably see that some of it chances to drop upon
the pistillate flowers. He may make similar observations with Indian corn and staminate pine
cones. A common agent in distributing pollen is
the wind. Plants which bear protruding feathery
stigmas and protruding stamens (as the grasses)
are generally wind-pollinated. So are many or
most diœcious or monœcious plants.

95. We have already referred to the fact (68)
that the showy petals sometimes attract insects.
The insects are also attracted by odors, as one may
infer by watching the visits of moths to the pe-
tunias at nightfall, at which time the flowers give
forth their odor. We would infer, therefore, that
those flowers which have neither showy colors nor
odors must be pollinated by the wind; and this
is true, as a general statement.

95a. Plants habitually pollinated by the wind are said to be
anemophilous ("wind-loving"), and those pollinated by insects entomophilous ("insect-loving").

95b. Since the publication of Darwin's remarkable investigations upon the inter-relations of flowers and insects, it has been commonly supposed that the showy colors of flowers have been developed, or ·have originated, as a means of attracting insects, but this explanation of the origin of colored parts is open to doubt. But whatever the evolution of the corolla may have been, it is known that color and perfume often attract insects.

96. It is evident that the insect would not visit the flower for the flower's sake, but for its own sake. There must be something in the flower which it wants, for color and odor are only attractions, not substantial rewards. The things which the insect wants are nectar (or honey) and pollen, chiefly the former.

FIG. 70.

Flower of columbine.

97. A flower of the columbine (often erroneously called honeysuckle) is shown in Fig. 70. The petals are produced into long spurs. If one of these spurs were opened when the flower is in full bloom, the bottom of it would be found to contain a glistening secretion. This is the nectar; and the spurs are, therefore, nectaries.

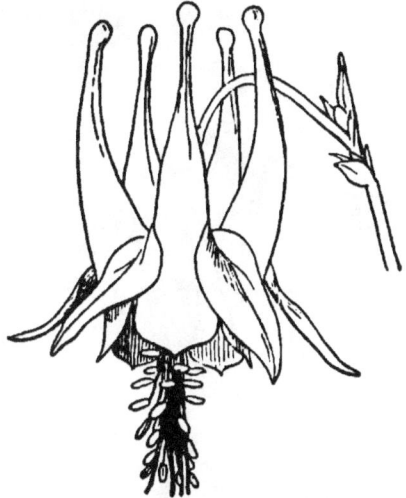

97a. Humming-birds are fond of sipping the nectar from the columbine, for which their long bills are eminently fitted. Bees

also crowd into the tubes. Bumble-bees often bite open the nectaries and steal the honey from the outside; this kind of theft is
not infrequent in other flowers.

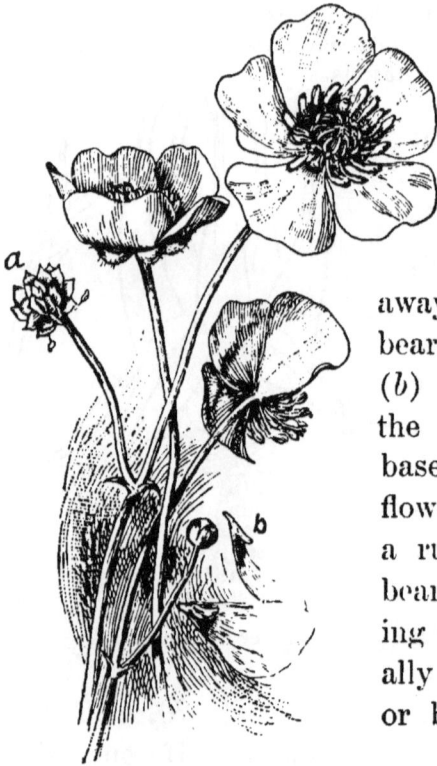

98. The pupil should
now examine any of
the buttercups, or
crowfoots. The com-
mon one in the East
is shown in Fig. 71.
If the petals are pulled
away, each one is seen to
bear a minute gland or lip
(*b*) at its base. This is
the nectary. The disk-like
base of the common grape
flower is also a nectary. As
a rule, entomophilous flowers
bear nectaries, or nectar-bear-
ing glands, and they are usually located in the very base
or bottom of the flower.

FIG. 71.

Flowers of common
buttercup.

SUGGESTIONS.—The pupil should now
look for the nectaries in all flowers
which he suspects to be insect-polli-
nated. The presence of spurs and
sacs, and also of glands, is presumptive evidence of nectaries.
The presence of insects about flowers always raises the presump-
tion that those flowers are entomophilous ; the pupil should, therefore, determine what visitors the common flowers may have.

IV. Propagation and Habits

XVI. HOW A SQUASH PLANT GETS OUT OF THE SEED

99. The culmination of the activities of the plant is the propagation of itself. To this end the devious life-history, the mechanisms of the flowers, the varieties and peculiarities of the fruits, are subordinate. The supreme effort of the plant — if one may so speak — is its perpetuation. The most important vehicle of this perpetuation, in most higher plants, is the seed.

100. If one were to plant seeds of a Hubbard or Boston Marrow squash in loose, warm earth in a pan or box, and were then to care for the parcel for a week or ten days,

FIG. 72.

Squash plant a week old.

FIG. 73.

Squash plant which has brought the seed - coats out of the ground.

(79)

he would be rewarded by a colony of plants like
that shown in Fig. 72.   If he had not planted
the seeds himself, or had
not seen such plants before,
he would not believe that
these curious plants would
ever grow into squash vines,

FIG. 74.

Germination just
beginning.

FIG. 75.

The root
and peg.

so different are they from the vines which we
know in the garden.   This, itself, is a most
interesting fact,— this wonderful difference between
the first and the later stages of all plants, and it
is only because we know it so well that we do
not wonder at it.

101. It may happen, however, that one or two
of the plants may look
like that shown in Fig.
73.   Here the seed
seems to have come up
on top of the plant,
and one is reminded of the curious
way in which beans come up on
the stalk of the young plant.
We are desirous to know why one
of these squash plants brings its
seed up out of the ground while
all the others do not.   We shall ask the plant.
We may first pull up the two plants.   The first
one (Fig. 72) will be seen to have the seed-coats

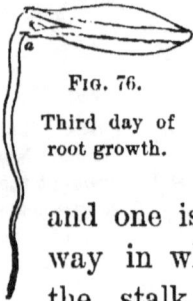

FIG. 76.

Third day of
root growth.

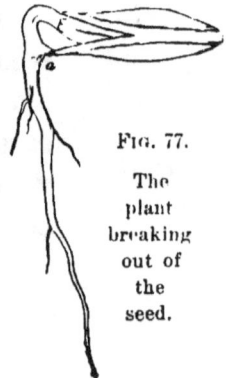

FIG. 77.

The
plant
breaking
out of
the
seed.

still attached to the very lowest part of the stalk, below the soil, but the other plant has no seed at that point.

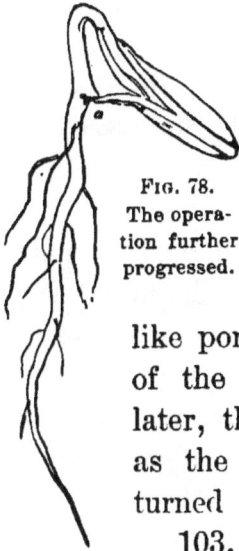

102. We will now plant more seeds—a dozen or more of them—so that we shall have enough for examination two or three times a day for several days. A day or two after the seeds are planted, we shall find a little point or root-like portion breaking out of the sharp end of the seed, as shown in Fig. 74. A day later, this portion has grown to be as long as the seed itself (Fig. 75), and it has turned directly downward into the soil.

FIG. 78. The operation further progressed.

103. There is another most curious thing about this germinating seed. Just where the root is breaking out of the seed (shown at *a* in Fig. 75), there is a little peg or projection. In Fig. 76, about a day later, the root has grown still longer, and this peg seems to be forcing the seed-coats apart. In Fig. 77, however, it will be seen that the seed-coats are really being forced apart by the stem or stalk above the peg, for this stem is now growing longer. The lower lobe of the seed has attached to the peg (seen at *a*, Fig. 77), and the

FIG. 79. The plant just coming up.

seed-leaves are backing out of the seed. Fig. 78 shows the seed a day later. The root has now produced many branches, and has thoroughly established itself in the soil. The top is also growing rapidly, and is still backing out of the seed, and the seed-coats are still firmly held by the obstinate peg.

FIG. 80.

The plant liberated from the seed-coats.

104. In the meantime, the plantlets which we have not disturbed have been coming through the soil. If we were to see the plant in Fig. 78 as it was "coming up," it would look like Fig. 79. It is tugging away trying to get its head out of the bonnet which is pegged down underneath the soil. In Fig. 80 it has escaped from its trap. It must now straighten itself up, as it is doing in Fig. 81, and it is soon standing stiff and straight, as in Fig. 72. We now see that the reason why the seed came up on the plantlet in Fig. 73, is because in some way the peg did not hold the seed-coats down (see Fig. 84), and the expanding leaves are pinched together, and they must get themselves loose as best they can.

FIG. 81.

The plant straightening up.

105. There is another thing about this squash plant which we must not fail to notice, and this

is the fact that these first two leaves came out of the
seed, and did not grow out of the plantlet itself. We
must notice, too, that these
leaves are much smaller
when they are first drawn
out of the seed than they are
when the plantlet has straightened itself up. That is, these leaves increase
in size after they reach the light and air.

106. The roots are now established
in the soil, and are taking in food
which enables the plantlet to grow.
The next leaves which appear (Fig.

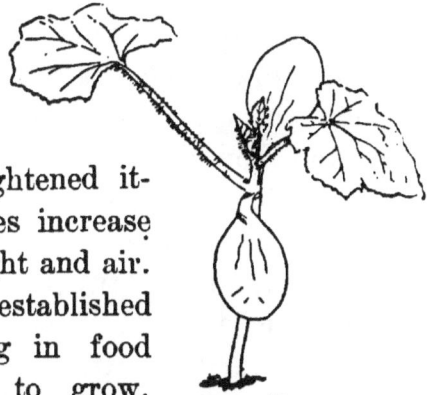

FIG. 82.
True leaves.

82) are very different from these first or seed
leaves. They grow out of the little plant itself.
The picture shows these true leaves as they appear
on a young Crookneck squash plant, and the plant
now begins to look much like a
squash vine.

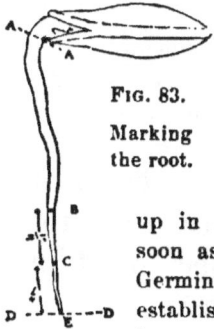

FIG. 83.
Marking
the root.

106a. The leaves which are borne in the
seed are the cotyledons or seed-leaves. Their
enlargement, after sprouting, is largely or wholly
at the expense of the nutriment which is stored
up in the seed. The true leaves (Fig. 82) appear as
soon as the plantlet begins to gather materials for itself.
Germination is not complete until the plantlet has thoroughly
established itself in the soil, and the true leaves have
begun to appear. The plantlet then becomes a plant.
The earlier part of the germinating process may be called sprouting.

106b. The incipient shoot which gives rise to the growth

above the cotyledons (and which, as we shall see, is present in the seed) is the plumule. The plumule is really a bud.

107. We are now curious to know how the stem grows when it backs out of the seed and pulls the little seed-leaves with it, and how the root grows downwards into the soil. Pull up another seed when it has sent a single root about two inches deep into the earth. Wash it very carefully and lay it upon a piece of paper. Then lay a rule alongside of it, and make an ink mark one-quarter of an inch, or less, from the tip, and two or three other marks at equal distances above (Fig. 83). Now carefully replant the seed. Two days later, dig it up; we shall most likely find a condition something like that in Fig. 84. It will be seen

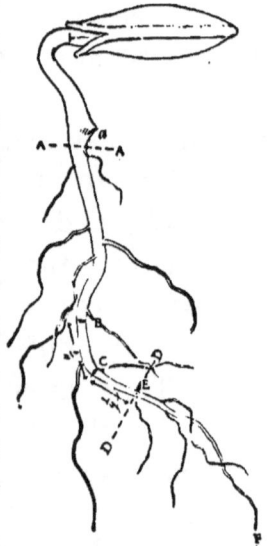

Fig. 84.

The root grows in the end portion.

that the marks E, C, B, are practically the same distance apart as before, and they are also the same distance from the peg, A A. The point of the root is no longer at D D, however, but has moved on to F. The root, therefore, has grown almost wholly in the end portion.

107*a*. Common ink will not answer for this purpose because it "runs" when the root is wet, but indelible ink, used for marking linen or for drawing, should be used. It should also be said that the roots of the common pumpkin and of the summer bush squashes are too fibrous and branchy for this test.

107*b*. It should be stated that the root does not grow at its very tip, but chiefly in a narrow zone just back of the tip; but the determination of this point is rather too difficult for the beginner, and, moreover, it is foreign to the purpose of this lesson.

108. Now let us make a similar experiment with the stem or stalk. Mark a young stem, as at A in Fig. 85, but the next day we shall find that these marks are farther apart than when we made them (B, Fig. 85). The marks have all raised themselves above the ground as the plant has grown. The stem, therefore, has grown throughout its length rather than from the end. The stem usually grows most rapidly, at any given time, in the upper or younger portion; but the part soon reaches the limit of its growth and becomes stationary, and the growth continues beyond it. (See "Suggestions," p. 23).

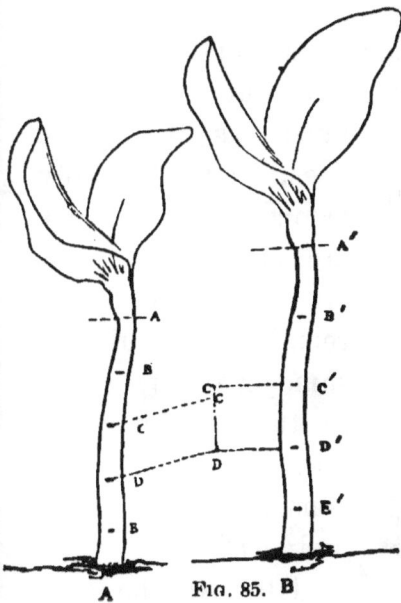

Fig. 85.

The marking of the stem and the spreading apart of the marks.

SUGGESTIONS.—All this behavior of the germinating squash results in raising the foliage above the soil and in keeping the seed-
coats beneath it. But suppose that the seed is not buried, but lies
on the surface of the moist earth, or is covered only with loose
leaves or litter: then what happens? Fill a pot or box with earth
up to half an inch below the rim, lay fresh squash seeds upon it,
cover the pot with cardboard and keep the seeds moist and warm.
Watch the result. Peas germinate in this way very readily.

XVII.   GERMINATION OF BEANS

109. Plant a few common beans and watch the
germination. The plantlets back out of the soil
much as the squash
does, and the coty-
ledons, *a*, Fig. 86,
are elevated into the
air. These cotyledons
remain practically the
same size as they were
in the seed, however,
and do not become
conspicuously green
and leaf-like.

FIG. 86.

Germination of
common bean.

FIG. 87

Germination of
Scarlet Runner bean.

110. At the same
time, plant seeds of
the Scarlet Runner or

White Dutch Runner bean.   The first foliar parts
to appear are true leaves (Fig. 87), and if the

plant be dug up, the cotyledons will be found to have remained under the ground. Observe carefully at what point the roots start out from the seed.

111. There are, then, two types of germination as respects the position of the cotyledons. In one type, the seed-leaves rise above ground, or the germination is epigeal ("above the earth"); in the other, they remain where the seed was planted, or the germination is hypogeal ("below the earth").

111a. The pupil should make a careful comparison of the differences in germination between the two types of beans mentioned above. He may profitably add a third factor to the experiment by including the garden pea. If he has access to oak trees, he may watch the germination of the acorns as they lie upon the ground in very early spring. Examine horse-chestnuts.

112. Measure the beans before they are planted, taking the length, width and thickness. If delicate balances or scales are at hand, it may be well to weigh them, also. Then observe the increase in size of the beans. Is this swelling associated with heat or moisture, or both? The pupil can answer this question by planting some seeds in dry, warm earth and others in moist, cool earth (which is kept little above freezing), and by otherwise varying the experiment. Do dead seeds— those which are very old or which have been baked—swell when planted? The pupil will find

that the swelling of the seed is the first obvious stage in germination.

113. Plant beans in moist cotton or sawdust, or lay them in folds of heavy, damp cloth. How far will the sprouting progress? Will the first true leaves develop? In other words, for how long can the plantlet grow upon the nutriment which is stored in the seed?

114. When the true leaves have begun to develop (as in Figs. 86 and 87), carefully lift the plant, with the soil which is attached, and then, in a basin, wash away the earth until the roots are white and clean. Then, by the aid of a lens or by holding the roots to the light, see the covering of very fine hairs upon the roots. It is these little organs which hold most of the earth on the roots when the plant is carefully pulled up. They are the root-hairs, and they are active agents in absorbing food.

114*a*. Several profitable lessons may be made in the study of the root-hairs of whatever seedling plants may be at hand in gardens or elsewhere. How soon after germination do they appear? Do they persist as the root becomes old, or are they shed upon the older parts? Do full-grown or large plants have root-hairs? Look for them on the very youngest parts of the roots. When seeds are germinated as recommended in 113 and p. 86, the root-hairs are much more readily seen.

115. We know that roots go downwards and stems go upwards. How soon is this difference

manifested in the germinating bean? Do the two parts take these opposite directions even when the beans germinate in a dark place? We shall find that there is an inherent, or inborn, tendency for the root to grow down and the stem to grow up.

115a. The discussion of the physiological causes which have determined this differentiation between the root and stem is not germane to this book; although it may be said that gravitation plays an important part in the movements. For the purpose of designating some of these facts or phenomena, the words geotropism and heliotropism have been used,—the former designating movement into or towards the earth, and the latter movement towards the light.

SUGGESTIONS.—The pupil should make these tests with beans. He will find other interesting points, if he watches the process of germination closely. When some of the seeds have produced straight roots an inch or so long, remove them carefully and hang them with the root uppermost in a moist and warm atmosphere, as under a bell-jar or inverted glass bowl which is set in water. Observe how the roots tend to turn downwards and the plumule to turn upwards. Or, sprouting seeds may be placed in a horizontal position. It is interesting to observe how the root gets around stones and other hard objects in the soil. The roots of any plant which grows in very stony or hard, gravelly soil are good subjects for observation. Radishes are also interesting for germinating studies; and they show heliotropism quickly and emphatically when growing where the light all comes from one direction, as from a side window.

The beans may be planted in pans or boxes which are set in the windows of the school-room, although care should be taken that the soil does not become too dry if it is exposed directly to the sun. Handframes, or bell-glasses, will- be found to be useful with which to make germination studies. The pupil usually takes more interest in the experiments, however, if he has them constantly under his eye. Perhaps each pupil can be provided with a small flower-pot, or other dish, or even with a cigar-box, and be allowed to have it upon his desk.   If the elongation of the parts is to be watched very closely, and especially if the root is to be marked in order to observe its method of growth (107, 107*a*), the seeds may be germinated between damp blotting papers. If a dozen or more seeds are started, a record may be kept of the various stages in germination by pressing the plants, as well as by drawing them.

XVIII. WHAT IS A SEED?

116. The two most important characteristics of seeds we have already learned,—the facts that they are the result of the fertilization of the ovule by a pollen grain, and that they con- tain a miniature plant. This condensed and miniature plant in the seed is called the embryo. The phenomena of fertilization are too obscure to be clearly understood, much less to be

Fig. 88.

The parts of a bean seed.

seen, by the beginner; but it may be said that the nucleus of the pollen-grain unites with the nucleus of the egg-cell in the embryo-sac, and the result of this union is the embryo.

117. Let us return to the bean. In the ripe pod the beans are seen to be attached by short stalks to the edge of each valve.   The

stalk is called, in all seeds, the funiculus. When the funiculus breaks away from the seed, it leaves a scar (D, Fig. 88). This scar is the hilum.

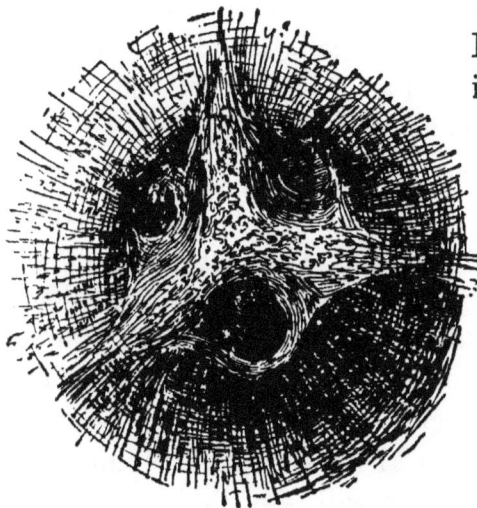

118. If we split the bean lengthwise (preferably after it has been soaked in water for a few hours), we find that the seed is composed of two thick cotyledons; and these are the parts which are afterwards elevated into the air (*a*, Fig. 86). One-half of a bean (that is, one cotyledon) is shown in Fig. 88. The other cotyledon was attached at C. The plumule is at S, and the incipient stem, or caulicle, at O. All these parts — cotyledons, caulicle, plumule, — constitute the embryo.

FIG. 89.

Micropylar scars of cocoa-nut.

118*a*. Over the point of the caulicle, the close observer will find a minute depression and a hole leading into the bean. This is the micropyle, and is the point at which the pollen-tube entered, and the place through which the root breaks in germination. In the cocoa-nut the positions of the three micropyles are shown by the scars (Fig. 89), but since only one of the locules develops a seed, germination takes place through only one place.

119. We are now curious to know if there is anything in the structure of the embryos to suggest the different behaviors of the two beans in Figs. 86 and 87. Fig. 90 shows the cotyledons of a common bean laid open; and Fig. 91 is a similar picture of the Scarlet Runner bean. The node, or place of attachment of the cotyledons, is at C (one cotyledon, of course, having been broken away). The caulicles are the parts pointing downwards, and the two leaves of the plumule lie at the left. The observer will see that the space between C and the plumule is very different in the two beans. These different lengths are suggestive of what takes place in germination,—the greatest elongation of the stem in the common bean takes place beneath the cotyledons, whereas the greatest elongation in the Scarlet Runner takes place above the cotyledons: in one case the caulicle elongates, in the other it does not.

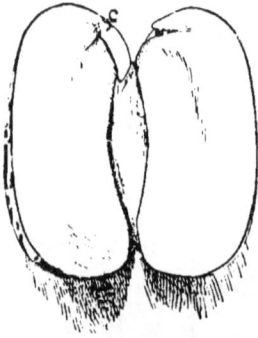

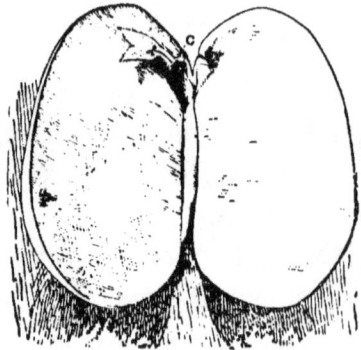

Fig. 90.

Parts of common bean.

Fig. 91.

Parts of Scarlet Runner bean.

119a. The internode lying above the point of attachment of the cotyledons—between the cotyledons and the plumule—is called the epicotyl; that below the cotyledons is the hypocotyl. The hypocotyl was formerly called the radicle, upon the supposition that it is an incipient root; but it is a stem (except, perhaps, the very tip), and the root develops from its end.

120. The embryos of the squash and bean oc-cupy the whole interior of the seed, and the

Fig. 92.

Section of
onion seed.

nutriment which sustains the sprouting plantlet is stored in the cotyledons. In the onion it is not so. Fig. 92 is a section of an onion seed. The monocoty-ledonous embryo is coiled up in a mass of starchy matter; and a similar condition is seen in the buckwheat (Fig. 93). Nutritive material stored outside the embryo is called the endosperm.

Fig. 93.

Section of
buckwheat
seed.

120a. The endosperm varies greatly in quantity and in physical and chemical character. In many plants, as the cereal grains, it affords most of the material which is utilized for human food. In the cocoa-nut only a part of the liquid matter solidifies into endosperm, leaving the "milk" in the center; the embryo is comparatively very small, and can be found, of course, near the micropyle.

121. A seed, then, is a body which is the direct product of a flower and a result of a sexual process, and which contains a miniature plant or embryo; and its office is to produce a new plant. In nearly all cases the embryo is

enclosed in seed-coats, and it is often imbedded in endosperm.

SUGGESTIONS.—Only the most obvious parts and features of seeds have been mentioned here, but these characters are sufficient to enable the pupil to make profitable comparisons between any seeds which may come to his hand. It is good practice to set beginners searching for the cotyledons in large seeds. Where, for example, are the cotyledons in the pecan (Fig. 94), acorn, maize,

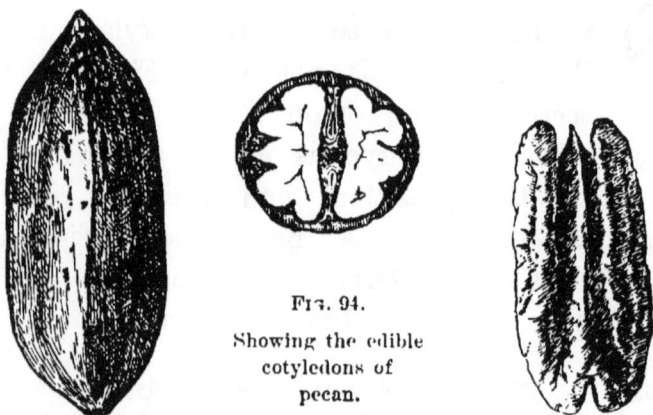

FIG. 94.

Showing the edible
cotyledons of
pecan.

wheat, castor bean, sweet pea, apple, peach, morning-glory, garden balsam, cucumber, orange, canna, cocoa-nut, and other large seeds? The parts can generally be made out more easily if the seeds are soaked in tepid water for a few hours. It is more important, however, if facilities are at hand, to set the pupil to the study of the behavior of seeds and plantlets in germination. An interesting series of studies can be made from a comparison of the form of the cotyledons with that of the first true leaves, and of the grada-tion from the character of the first leaves to those which are characteristic of the mature plant. Are there differences in coty-ledons and first leaves between the different horticultural varieties of the same plant, as between the different kinds of tomatoes.

XIX. BULBS, BULBLETS AND BUDS

122. We have now found that plants propagate themselves by both sexual and sexless means, for seeds are the product of a sexual process, whereas spores are not.

123. A bulb of an

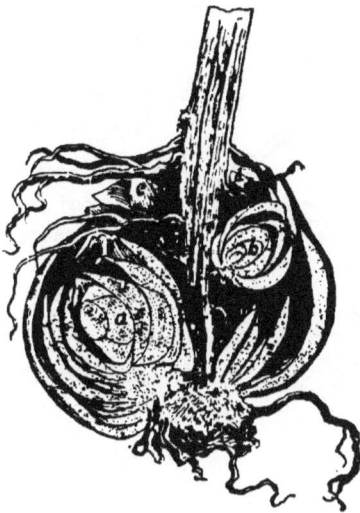

FIG. 95.

Section showing the formation of the bulbels.

FIG. 96.

Onion bulbs.

Easter lily is shown in Fig. 95. It is breaking up into several parts; and the gardener knows that each of these parts becomes a new bulb.

124. If we cut the bulb, we find a main

axis, and separate bulbs (or bulbels) are forming at
a, *b*, *c*, *d*. Each of these bulbels, as well as the
mother-bulb, is seen to be only a mass of thickened
scales, and these scales are
transformed leaves. A bulb,
then, is only a special kind
of bud.

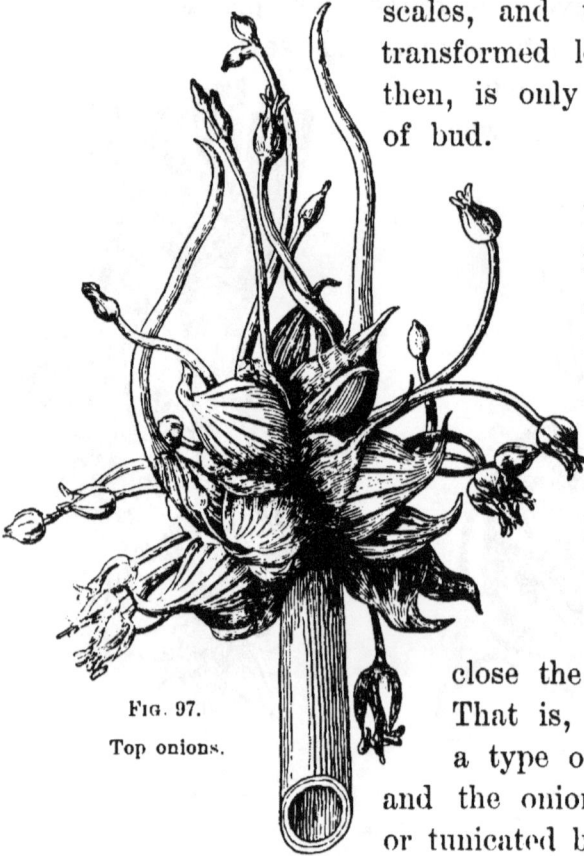

125. Onion
bulbs are shown
in Fig. 96. They
are of a different
make-up from
the lily bulb, for
the parts, in-
stead of being
narrow and over-
lapping longitu-
dinally, are thin
plates which en-
close the interior plates.
That is, the lily bulb is
a type of a scaly bulb,
and the onion of a laminate
or tunicated bulb.

Fig. 97.

Top onions.

125*a*. One of these onions has a very thick neck or stalk and a
comparatively small bulb. The top has grown at the expense of the
bottom, and the bulb is worthless for market. Such onions are known
as scullions.

126. The onion produces flowers in umbels.
Fig. 97 is a bunch of "top onions," in which
bulbs (or bulblets) are
borne in the flower-
cluster. If the pupil
examines such a
cluster he may find,
as in this picture,
an umbel bearing
flowers, well-formed
bulblets, and leaves
springing from im-
perfect or scullion-
like bulblets. In

Fig. 98.

Rosette and offsets of house-leek.

other words, flowers have been tranformed into purely
vegetative parts.

126a. The pupil may have access to the tiger lily, which bears
bulblets in the axils of the leaves. Top onions may be had of any
seedsman.

127. If bulbs are buds, then we should expect
to find various intermediate forms. The house-leek
(better known as hen-and-chickens, old-man-and
woman) produces dense rosettes of leaves on the
ground (Fig. 98). This rosette is structurally a
loose, open-topped bulb. The young rosettes, or
offsets, are produced upon short stems from the
under side of the rosette, rather than by the

growth of interior parts, as in the lily; but there are some true bulbs which propagate in a similar way.

127*a*. Let the pupil examine the bulbs of the dog's-tooth violet or "adder's-tongue,"—which gladdens the copses with its nodding bell-like flowers in earliest spring,— for a method of propagation comparable with that of the house-leek.

128. A head of cabbage is cut in two in Fig. 99. It is made up of overlapping and thick-

Fig. 99.
Section of cabbage.

ened leaves, and is really a gigantic bud. There is this important difference between the cabbage and a lily bulb and house-leek rosette, however, that the cabbage bud is not a means of propagating the plant, and one head or bud does not give rise directly to another. It is simply a storehouse; and in this case, the bud has been developed by man through the process of continually selecting for seed plants which have the densest or most coveted buds or heads.

129. We can distinguish bulbs from normal buds, then, by saying that bulbs directly give rise to other bulbs which produce plants; and these plants may produce bulbs directly, or may bear

FIG. 100.
Winter bud of anacharis.

FIG. 101.
Winter bud of myriophyllum.

seeds which produce plants which produce bulbs. Buds give rise to growing shoots which may produce flowers and seeds, and these seeds produce plants which produce buds. We cannot carry this distinction far, however, because bulbs not only produce other bulbs by lateral growth, but at the same time produce a growing vertical shoot or axis; and we shall find, also, that buds may separate from the parent in essentially the same way that the bulblets of the tiger lily do. The point is that plants may propagate by either sexual or asexual means, or by both means.

130. If one were to pull the water-weeds from the drift on the margins of lakes and ponds in late fall, he would find many of the strands with large bud-like bodies at the ends (Figs. 100, 101). These buds drop to the bottom of the pond, and in spring vegetate and give rise to new plants.

SUGGESTIONS.—Horticulturists raise onions in four ways: by sowing the seed; by planting bulblets (Fig. 97); by "multipliers," which are bulbs that break up into several bulbs during the process of growth; by sets, which are small bulbs that have been purposely arrested in their growth the previous year (by sowing seed in dry ground and allowing the plants to stand very close together) and which, when planted, complete their growth and become merchantable bulbs.

XX. HOW SOME PLANTS GET UP IN THE WORLD

131. The hop reaches light and air by coiling around some support (Fig. 102). If the pupil has access to a hop-field (hops often grow on old fences) or to the Japanese hop of gardens, let him observe the direction in which the stems twine. He will find the tips coiling from his right to his left, or in the direction of the sun's movement.

132. The morning-glory (Fig. 103) twines in the opposite direction,—from the observer's left to right. Fig. 104 is a morning-glory shoot which was taken from its support, and the free end,— above the string,— coiled about the stake in the

opposite direction. Two hours thereafter, the shoot
had uncoiled itself and the tip, as seen in the
picture, was again resuming its natural direction.

Fig. 102.	Fig. 103.	Fig. 104.
Japanese hop,—with the sun.	Morning-glory,—against the sun.	Morning-glory refusing to twine with the sun.

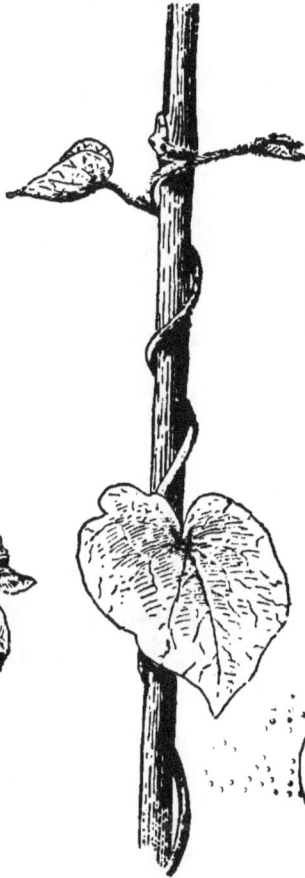

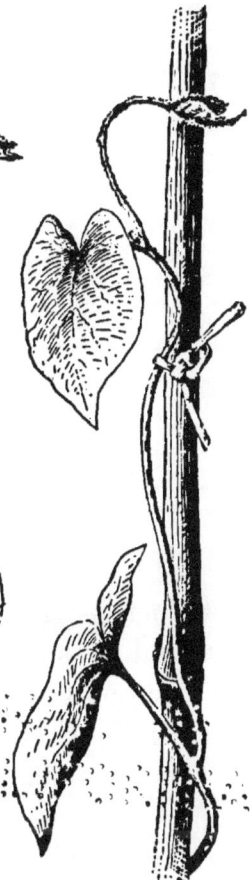

We shall expect to find that most kinds of twining plants coil in only one direction.

132*a*. Plants which coil with the sun, or from the observer's
right to left, are known as sinistrorse or eutropic ; those which coil

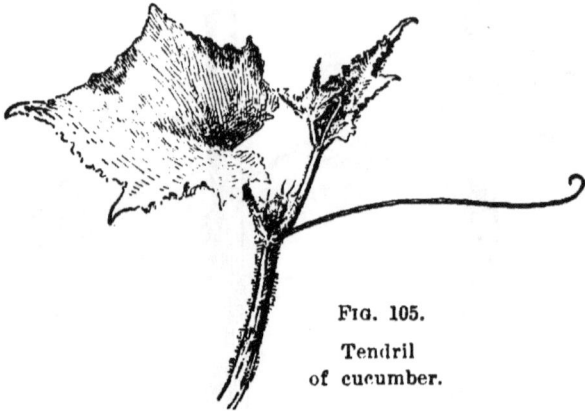

Fig. 105.
Tendril
of cucumber.

against the sun, or left to right, are dextrorse or antitropic. The latter direction is the more common.

133. Let the pupil watch the free end of a
twiner, — as on a young plant which has not yet
found a support, or a long tip projecting above
a support — and take note of the position or direction of the tip at different times of the day.
He will find that the tip revolves in a plane, as
if seeking a support.

134. The cucumber climbs by means of tendrils (Fig. 105). Notice that the tendril is hooked,
in readiness to catch a support. Does the point
of the tendril revolve? Watch it closely; or draw

a mark along one side of it, from base to tip,
with indelible ink, and observe if the line be-

FIG. 106.

Tendrils of cassabanana, a melon-like plant.

comes twisted, or if it is now seen on the con-
cave side of the tendril and then on the con-
vex side.

135. The tendril finally strikes a support.
What then? It coils; but if it coils much, why
does it not twist in two, since both base and
tip are fixed? (Study Fig. 106. At *a* the branches
of the tendril are searching for a support. At *b*

two of the branches have found support, and have coiled spirally, thereby drawing the plant near the support; but notice that there are places in each where one coil is missing. At these places, the direction of the coil was changed. The middle branch failed *at times* to find a support, and has twisted up into a querl.; and the same thing has occurred in *c.*)

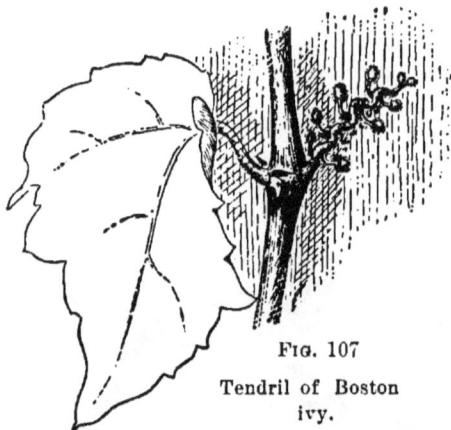

Fig. 107

Tendril of Boston ivy.

135*a.* Farmers' boys say that a watermelon is ripe when the querl is dead (which, however, may not be true). What is this querl?

135*b.* The tendrils of some plants are provided with discs at the ends, rather than hooks, by means of which they attach to a support. Compare the common Virginia creeper; also the root-like tendrils of the Japanese ampelopsis or Boston ivy (Fig. 107). Can the pupil show that the tendril in Fig. 107. is stem, not root?

136. A clematis is shown in Fig. 108. Here the petiolule of the terminal leaflet is acting as a tendril, although all of the petiolules and the petiole have the same habit. Leaves, then, may act both as tendrils and foliage.

136a. This recalls the fact that there are various disguises of leaves. Leaflets may be represented by tendrils. If the pupil will study the position of tendrils of the grape, he will find that they occupy the places of flower-clusters. (Has he not seen a bunch of grapes with one or two tendrils protruding?) Let him determine the morphology of the tendrils of cucumbers and melons. Observe, also, how the garden nasturtium, or tropæolum, climbs.

137. The trumpet creeper, poison ivy, true or English ivy, and some other plants, climb by roots which attach themselves to the support. Observe that such roots prefer to occupy the dark places or chinks on the building or bark upon which they climb.

FIG. 108.

Clematis climbing by leaf-tendril.

138. Some plants are mere scramblers, as some tall forms of blackberries, the galiums, some of the smart-weed tribe or polygonums. Such plants are often provided with various hooks or prickles by means of which they are secured to the support as they grow; but it by no means follows that all hooks or prickles on plants serve

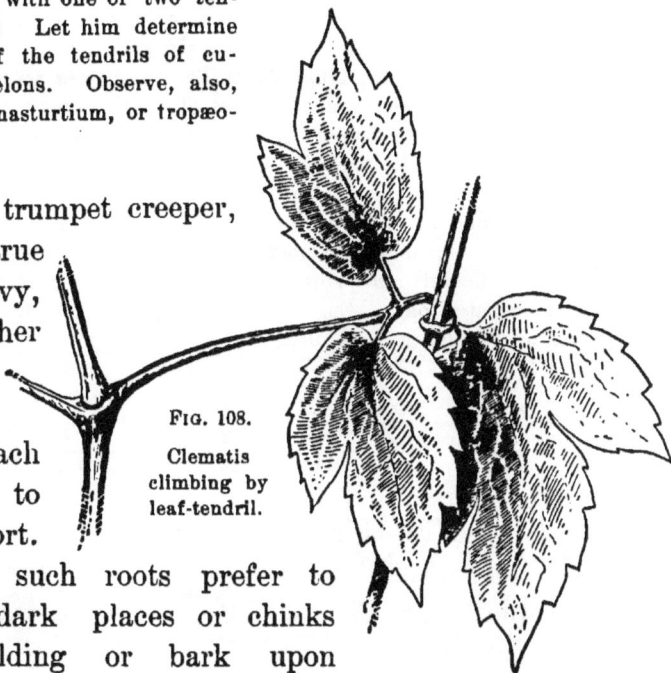

such a purpose, or, in fact, that they were developed primarily as a means of enabling the plant to climb.

SUGGESTIONS.—We have thus seen how some plants are able to maintain themselves in the fierce struggle for existence. Let the pupil observe if climbing plants naturally grow with other and tall plants, or do they frequent places of less competition and run their chances of finding support on other things than growing plants. Does the climbing habit impress the pupil as being a means of enabling the plant to reach light and air? In respect to the methods by which plants climb, any climber will afford interesting study, but the teacher will find young morning-glory, pea, pole bean, Japanese hop, cucumber, and nasturtium plants to be easily grown from seeds and useful in demonstration. Darwin's "Movements and Habits of Climbing Plants" should be consulted..

XXI.   VARIOUS MOVEMENTS OF PLANTS

139. With Fig. 26 we studied the form of the leaf of bean, but there is more to be seen in the picture. The leaf at the left was drawn in the day-time, that at the right in the night-time. There are similar differences in the positions of leaflets of oxalis (Figs. 109, 110) or wood-sorrel. Observe, also, at day and night, the leaves of clovers, lupines, locusts and acacias. In other words, the leaflets and leaves of many plants, notably of the Leguminosæ, take different positions at day and at night. The leaves of some plants close up during very hot hours of the day. The

leaves of purslane, and even of Indian corn and grasses, seem to wilt or to roll up when the weather is hot, and loss of moisture is thereby prevented.

140. The flower of the California poppy, or eschscholtzia, which is common in gardens, opens at day and closes at night. Observe, also, the flower of "pussley", the garden portulaca or rose-moss, oxalis, and some of the mallows. Other flowers open at night and close at day. This diurnal movement of the parts of plants is known as the "sleep of plants."

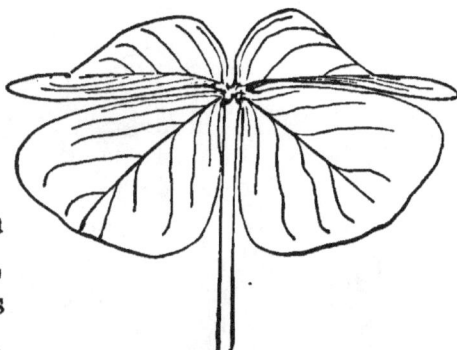

Fig. 109.

Day position of oxalis leaflets.

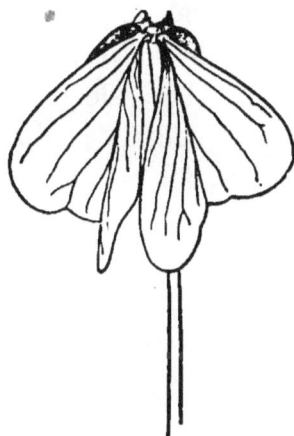

Fig. 110.

Night position of oxalis leaflets.

140a. It is not a sleep, however, in the sense of being a rest or period of recuperation for the plant. How these movements are produced is not definitely known, but they are associated intimately with the stimuli exerted by light and darkness, heat and cold. The utility of the movement is also in dispute. Darwin found that the position of sleeping leaves at night is such as to conserve the vital heat of the plant, and it is possible that some of this leaf-movement has arisen as a direct means of adaptation to circumstances or as a protection to the plant; but in the present state of our knowledge, this is largely assumption.

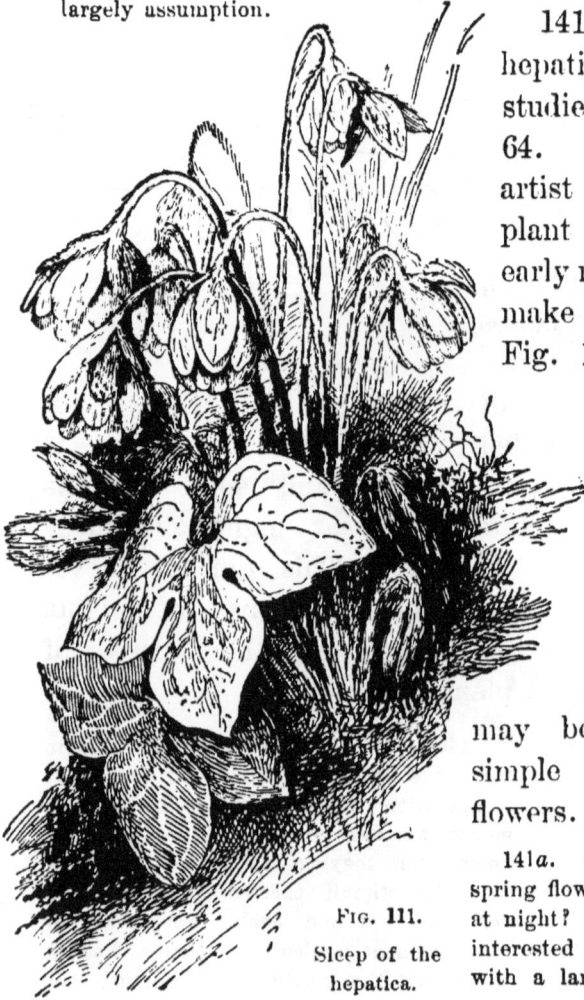

141. The flowers of hepatica have been studied in Fig. 42 and 64. If, however, the artist were to draw the plant at night or in early morning, he would make a picture like Fig. 111. The entire flower droops by the bending of the scape, and it straightens up and expands in the day-time. The sleep of plants, then, may be more than a simple closing of the flowers.

Fig. 111.

Sleep of the hepatica.

141a. Is it common for early spring flowers to close or to droop at night? The pupil may now be interested to explore the garden with a lantern.

142. One of the most remarkable movements in plants is that of the leaf and leaflets of the

sensitive plant (Fig. 112). The normal position of the leaf is shown at the right. A slight touch or shock causes the petiole to drop and the leaflets to shut up, as shown on the left. The movements are rapid and striking.

142*a*. The sensitive plant (Mimosa pudica) is easily grown from seeds, which may be obtained of seedsmen. It thrives wher-

FIG. 112.

The curious behavior of the sensitive plant.

ever beans will grow. The young plants, which grow rapidly, are more sensitive than old ones. The sensitive plant is one of the Leguminosæ.

143. We have now seen movements in stamens (Fig. 61), in leaves, the opening and closing of flowers, the shoots of twining plants and of tendrils, the fly-catchers of insectivorous plants, of stems towards light, and roots towards the earth (Obs. xvii.) and darkness (137). There are movements in the bursting of pods; and there are other movements which we have not considered. Plants are not as fixed and as unresponsive to external conditions as we have thought them to be.

V. Collecting

XXII. THE PRESERVING OF PLANTS

144. More than 100,000 species of flowering plants are known, and it is probable that nearly as many more await discovery. It is evident that if this vast number of facts is to be studied, the facts must be arranged or classified.

145. If the kinds of plants are to be carefully studied, specimens must be preserved. The plants of an entire region can then be seen, and, what is more important, they can be seen side by side, for comparative study is the only productive method in systematic or descriptive botany.

146. The plants are preserved by drying them under pressure. These dried and pressed plants are then secured to sheets of large white stiff paper (Fig. 113), and the sheets are filed away in covers, as leaves of music are placed in a portfolio. The covers are laid flat in a cupboard or cabinet. Such a collection of plants is an herbarium.

147. Although the specimens shrink some in drying and flowers often lose their color, these

dried plants preserve their distinctive characters remarkably well. It is from such specimens that

FIG. 113.

An herbarium sheet.

most descriptions of plants are made; and a person who proposes a new species always preserves a specimen of it as a record. In case of doubt as to what the species is, the specimen, rather than the description, is consulted.

148. A label (Fig. 114) should always accompany the specimen, and be securely glued to the sheet. The size, form and style of label are

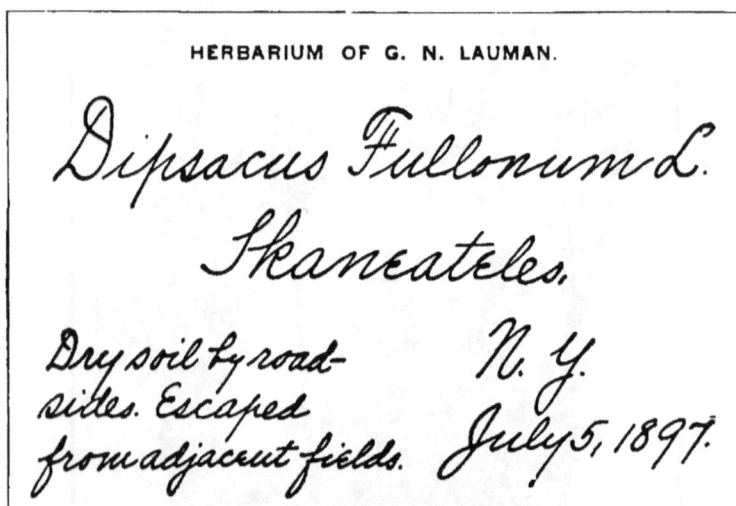

Fig. 114.

An herbarium label.

governed by the wishes of the maker of the herbarium ; but the label should give the name of the plant, where and when collected, and any incidental information, as to soil, location, color of flowers, height of plant, which is likely to be useful.

149. The collecting of the plants is botanizing. The first requisite is a tin case or vasculum (Fig. 115), in which the plants are placed, as collected. The specimens are pressed when the collector arrives home. If the vasculum closes tight, the specimens will remain in good condition for several hours. If they wilt too rapidly they may be lightly sprinkled with water. Upon journeys or long tramps, a portable press is sometimes used (shown in Fig. 115), the pressure being applied by means of straps. The most important point to be considered in collecting plants is to make sure that the specimen is large enough and good enough to fairly represent the plant from which it is taken. A good specimen is one which is well pressed and which comprises leaves, flowers and fruit; and a complete specimen is one which represents every part of the plant, including the root.

FIG. 115.

Collecting outfit.

150. It is important to remember that common plants are most useful for study, and several specimens should be taken, representing different soils and conditions. If one begins with the thought of securing only the rare, curious or beautiful things, he will probably have an herbarium which is of no particular value. He will have only a collection of detached plants. Some theme or motive should run through a collection,—to exhibit the flora of a neighborhood or a roadside, to illustrate the plants of a forest or a garden, to show the effects of different environments, and the like.

150*a*. In collecting plants, always set out with the ambition to make good specimens. Collect samples of all parts of the plant,—lower and upper leaves, stem, flowers, fruit, and, wherever practicable, roots. In small species, those two feet high or less, the whole plant should be taken. Of larger plants, take portions about a foot long. Press the plants between papers or "driers." These driers may be any thick porous paper, as blotting-paper or carpet-paper, or, for plants that are not succulent or very juicy, newspapers in several thicknesses may be used. It is best to place the specimens in sheets of thin paper—grocer's tea-paper is good—and place these sheets between the driers. Many specimens can be placed in a pile. On top of the pile place a short board and a weight of thirty or forty pounds, or a lighter weight if the pile is small and the plants are soft. Change the driers every day. The plants are dry when they become brittle, and when no moisture can be felt by the fingers. Some plants will dry in two or three days, while others require as many weeks. If the pressing is properly done, the specimens will come out smooth and flat, and the leaves will usually be green, although some ˙ plants always turn black in drying.

150*b*. Specimens are usually mounted on single sheets of white paper of the stiffness of very heavy writing-paper or thin Bristol·board. The standard size of sheet is $11\frac{1}{2}$ x $16\frac{1}{2}$ inches. The plants may be pasted down permanently and entirely to the sheet, or they may be held on by strips of gummed paper (as in Fig. 436). In the former case, Dennison's fish-glue is a good material to use. · Only one species or variety should be placed on a sheet. Specimens which are taller than the length of a sheet should be doubled over when they are pressed. The species of a genus are collected into a genus cover. This cover is a folded sheet of heavy manila or other firm paper, and the standard size, when folded, is 12 x $16\frac{1}{2}$ inches. On the lower left-hand corner of this cover the name of the genus is written. The specimens are now ready to be filed away. If insects attack the specimens, they may be destroyed by fumes of bisulphide of carbon (which is very inflammable) or chloroform. In this case it is necessary to place the specimens in a tight box and then insert the liquid. Lumps of camphor placed in the cabinet are useful in keeping away insects. Those who wish detailed information on the collecting of plants should consult W. W. Bailey's "Botanical Collector's Handbook." For methods of making leaf prints and of preserving flowers in natural colors (and of collecting and preserving insects), consult Chap. XV. of Bailey's "Horticulturist's Rule-Book," 4th edition.

150*c*. The naming of the specimens must be accomplished with the aid of some manual of the plants of the region. There are several books to aid in this work; but the teacher should bear in mind the important fact that the name of a plant is less important than the plant itself, and effort should not be expended in this direction at the expense of the study of the specimens. By making herbaria of the various forms of common species of plants, much of the labor of mere identification is avoided. The name of a plant serves two purposes: it affords language which we can use in speaking or writing of the plant, and it serves as an index to whatever may have been written about the plant.

150*d*. The standard systematic work upon the plants of North America is Gray's "Synoptical Flora," which, however, is not yet completed. For that part of the United States east of the Mississippi and north of Tennessee, and practically including adjacent Canada, Gray's "Manual," now in its sixth edition, is the standard authority. Britton and Brown's new "Illustrated Flora," in three volumes and with an illustration of every species, covers essentially the same territory as the manual, with the addition of the British Possessions as far north as Newfoundland. Macoun's "Catalogue of Canadian Plants," in several parts, and published by the Geological and Natural History Survey of Canada, may be consulted for the British Possessions. For the southern states east of the Mississippi, the third edition of Chapman's "Flora of the Southern States" is the standard reference. For the territory west of the Mississippi there is no single manual. The floras covering parts of this region are: Coulter's "Manual of the Botany of the Rocky Mountain Region," and "Flora of Western Texas," the latter published by the United States Department of Agriculture; Greene's "Manual of the Botany of the Region of San Francisco Bay," for central California; Howell's "Flora of Northwest America," for Oregon, Washington and Idaho. For the common wild and cultivated plants of the United States east of the Mississippi, the revision of Gray's "Field, Forest and Garden Botany" should be consulted. Books of a more popular nature may often be used by teacher or pupils, as Mrs. Dana's "How to Know the Wild Flowers," Mathews' "Familiar Flowers of Field and Garden," W. W. Bailey's "Among Rhode Island Wild Flowers," Baldwin's "Orchids of New England," Newhall's books upon "Trees," "Shrubs," and "Vines," Knobel's Guides ("Trees and Shrubs," "Ferns and Evergreens" of New England), and others. There are many excellent local floras, — books devoted to the plants of a state, county, or small circumscribed geographical area. Other systematic books are mentioned in "Lessons with Plants."

SUGGESTIONS. The collecting of natural objects is one of the delights of youth. Its interest lies not only in the securing of the objects themselves, but it appeals to the desire for adventure and exploration. Botanizing should be encouraged; yet there are cautions to be observed. The herbarium should be a means, not an end. To have collected and mounted a hundred plants is no merit; but to have collected ten plants which represent some theme or problem is eminently useful. Schools usually require that the pupils make an herbarium of a given number of specimens, but this is scarcely worth the effort. Let the teacher set each collector a problem. One pupil may make an herbarium representing all the plants of a given swale, or fence-row, or garden; another may endeavor to show all the forms or variations of the dandelion, pigweed, apple tree, timothy, or red clover; another may collect all the plants on his father's farm, or all the weeds in a given field; another may present an herbarium showing all the forest trees or all the kinds of fruit trees of the neighborhood; and so on. The collector should be asked to display his herbarium to the school, explaining the problem in hand; and the teacher and others may then criticise the making of the specimens. The teacher should discourage the collection of plants simply because they are rare; and an effort should be made to preserve in their natural locations the interesting and showy wild flowers, rather than to destroy them by over-zealous collecting.

Fig. 116.

The botanist's resort on a rainy day.

The Best and Newest Rural Books

BOOKS ON LEADING TOPICS CONNECTED WITH AGRICULTURAL AND RURAL LIFE ARE HERE MENTIONED. EACH BOOK IS THE WORK OF A SPECIALIST, UNDER THE EDITORIAL SUPERVISION OF PROFESSOR L. H. BAILEY, OF THE CORNELL UNIVERSITY, OR BY PROFESSOR BAILEY HIMSELF, AND IS READABLE, CLEAR-CUT AND PRACTICAL.

THE RURAL SCIENCE SERIES

Includes books which state the underlying principles of agriculture in plain language. They are suitable for consultation alike by the amateur or professional tiller of the soil, the scientist or the student, and are freely illustrated and finely made.

The following volumes are now ready:

THE SOIL. By F. H. KING, of the University of Wisconsin. 303 pp. 45 illustrations. 75 cents.

THE FERTILITY OF THE LAND. By I. P. ROBERTS, of Cornell University. 421 pp. 45 illustrations. $1.25.

THE SPRAYING OF PLANTS. By E. G. LODEMAN, late of Cornell University. 399 pp. 92 illustrations. $1.00.

MILK AND ITS PRODUCTS. By H. H. WING, of Cornell University. 311 pp. 43 illustrations. $1.00.

THE PRINCIPLES OF FRUIT-GROWING. By L. H. BAILEY. 516 pp. 120 illustrations. $1.25.

BUSH-FRUITS. By F. W. CARD, of Rhode Island College of Agriculture and Mechanic Arts. 537 pp. 113 illustrations. $1.50.

FERTILIZERS. By E. B. VOORHEES, of New Jersey Experiment Station. 332 pp. $1.00.

THE PRINCIPLES OF AGRICULTURE. By L. H. BAILEY. 300 pp. 92 illustrations. $1.25.

IRRIGATION AND DRAINAGE. By F. H. KING, University of Wisconsin. 502 pp. 163 illustrations. $1.50.

THE FARMSTEAD. By I. P. ROBERTS. 350 pp. 138 illustrations. $1.25.

RURAL WEALTH AND WELFARE. By GEORGE T. FAIRCHILD, Ex-President of the Agricultural College of Kansas. 381 pp. 14 charts. $1.25.

THE PRINCIPLES OF VEGETABLE-GARDENING. By L. H. BAILEY. 468 pp. 144 illustrations. $1.25.

THE FEEDING OF ANIMALS. By W. H. JORDAN, of New York State Experiment Station. 450 pp. $1.25 net.

FARM POULTRY. By GEORGE C. WATSON, of Pennsylvania State College. 341 pp. $1.25 net.

THE FARMER'S BUSINESS HANDBOOK. By I. P. ROBERTS, of Cornell University. 300 pp. $1.00 net.

THE CARE OF ANIMALS. By NELSON S. MAYO, of Kansas State Agricultural College. 458 pp. $1.25 net.

THE HORSE. By I. P. ROBERTS, of Cornell University. 413 pp. $1.25 net.

New volumes will be added from time to time to the RURAL SCIENCE SERIES. The following are in preparation:

PHYSIOLOGY OF PLANTS. By J. C. ARTHUR, Purdue University.

THE PRINCIPLES OF STOCK BREEDING. By W. H. BREWER, of Yale University.

PLANT PATHOLOGY. By B. T. GALLOWAY and associates, of U. S. Department of Agriculture.

THE POME FRUITS (Apples, Pears, Quinces). By L. H. BAILEY.

THE GARDEN-CRAFT SERIES

Comprises practical handbooks for the horticulturist, explaining and illustrating in detail the various important methods which experience has demonstrated to be the most satisfactory. They may be called manuals of practice, and though all are prepared by Professor BAILEY, of Cornell University, they include the opinions and methods of successful specialists in many lines, thus combining the results of the observations and experiences of numerous students in this and other lands. They are written in the clear, strong, concise English and in the entertaining style which characterize the author. The volumes are compact, uniform in style, clearly printed, and illustrated as the subject demands. They are of convenient shape for the pocket, and are substantially bound in flexible green cloth.

THE SURVIVAL OF THE UNLIKE:
A Collection of Evolution Essays Suggested by the Study of Domestic Plants. By L. H. BAILEY, Professor of Horticulture in the Cornell University.

FOURTH EDITION — 515 PAGES — 22 ILLUSTRATIONS — $2.00

To those interested in the underlying philosophy of plant life, this volume, written in a most entertaining style, and fully illustrated, will prove welcome. It treats of the modification of plants under cultivation upon the evolution theory, and its attitude on this interesting subject is characterized by the author's well-known originality and independence of thought. Incidentally, there is stated much that will be valuable and suggestive to the working horticulturist, as well as to the man or woman impelled by a love of nature to horticultural pursuits. It may well be called, indeed, a philosophy of horticulture, in which all interested may find inspiration and instruction.

THE SURVIVAL OF THE UNLIKE comprises thirty essays touching upon The General Fact and Philosophy of Evolution (The Plant Individual, Experimental Evolution, Coxey's Army and the Russian Thistle, Recent Progress, etc.); Expounding the Fact and Causes of Variation (The Supposed Correlations of Quality in Fruits, Natural History of Synonyms, Reflective Impressions, Relation of Seed-bearing to Cultivation, Variation after Birth, Relation between American and Eastern Asian Fruits, Horticultural Geography, Problems of Climate and Plants, American Fruits, Acclimatization, Sex in Fruits, Novelties, Promising Varieties, etc.); and Tracing the Evolution of Particular Types of Plants (the Cultivated Strawberry, Battle of the Plums, Grapes, Progress of the Carnation, Petunia, The Garden Tomato, etc.).

THE EVOLUTION OF OUR NATIVE FRUITS. By L. H. BAILEY, Professor of Horticulture in the Cornell University.

472 PAGES — 125 ILLUSTRATIONS — $2.00

In this entertaining volume, the origin and de velopment of the fruits peculiar to North America are inquired into, and the personality of those horticultural pioneers whose almost forgotten labors have given us our most valuable fruits is touched upon. There has been careful research into the history of the various fruits, including inspection of the records of the great European botanists who have given attention to American economic botany. The conclusions reached, the information presented, and the suggestions as to future developments, cannot but be valuable to any thoughtful fruit-grower, while the terse style of the author is at its best in his treatment of the subject.

THE EVOLUTION OF OUR NATIVE FRUITS discusses The Rise of the American Grape (North America a Natural Vineland, Attempts to Cultivate the European Grape, The Experiments of the Dufours, The Branch of Promise, John Adlum and the Catawba, Rise of Commercial Viticulture, Why Did the Early Vine Experiments Fail ! Synopsis of the American Grapes); The Strange History of the Mulberries (The Early Silk Industry, The "Multicaulis Craze,"); Evolution of American Plums and Cherries (Native Plums in General, The Chickasaw, Hortulana, Marianna and Beach Plum Groups, Pacific Coast Plum, Various Other Types of Plums, Native Cherries, Dwarf Cherry Group); Native Apples (Indigenous Species, Amelioration has begun); Origin of American Raspberry-growing (Early American History, Present Types, Outlying Types); Evolution of Blackberry and Dewberry Culture (The High-bush Blackberry and Its Kin, The Dewberries, Botanical Names); Various Types of Berry-like Fruits (The Gooseberry, Native Currants, Juneberry, Buffalo Berry, Elderberry, High-bush Cranberry, Cranberry, Strawberry); Various Types of Tree Fruits (Persimmon, Custard-Apple Tribe, Thorn-Apples, Nut-Fruits); General Remarks on the Improvement of our Native Fruits (What Has Been Done, What Probably Should Be Done).

L ESSONS WITH PLANTS: Suggestions for Seeing and Interpreting Some of the Common Forms of Vegetation. By L. H. BAILEY, Professor of Horticulture in the Cornell University, with delineations from nature by W. S. HOLDSWORTH, of the Agricultural College of Michigan.

SECOND EDITION—431 PAGES—440 ILLUSTRATIONS—12 MO— CLOTH—$1.10 NET

There are two ways of looking at nature. The *old way*, which you have found so unsatisfactory, was to classify everything—to consider leaves, roots, and whole plants as formal herbarium specimens, forgetting that each had its own story of growth and development, struggle and success, to tell. Nothing stifles a natural love for plants more effectually than that old way.

The new way is to watch the life of every growing thing, to look upon each plant as a living creature, whose life is a story as fascinating as the story of any favorite hero. "Lessons with Plants" is a book of stories, or rather, a book of plays, for we can see each chapter acted out if we take the trouble to *look* at the actors.

"I have spent some time in most delightful examination of it, and the longer I look, the better I like it. I find it not only full of interest, but eminently suggestive. I know of no book which begins to do so much to open the eyes of the student—whether pupil or teacher—to the wealth of meaning contained in simple plant forms. Above all else, it seems to be full of suggestions that help one to learn the language of plants, so they may talk to him."—DARWIN L. BARDWELL, *Superintendent of Schools, Binghamton.*

"It is an admirable book, and cannot fail both to awaken interest in the subject, and to serve as a helpful and reliable guide to young students of plant life. It will, I think, fill an important place in secondary schools, and comes at an opportune time, when helps of this kind are needed and eagerly sought."—Professor V. M. SPALDING, *University of Michigan.*

FIRST LESSONS WITH PLANTS

An Abridgement of the above. 117 pages—116 illustrations—40 cents net.

BOTANY: An Elementary Text for Schools. By L. H. BAILEY.

355 PAGES—500 ILLUSTRATIONS—$1.10 NET

"This book is made for the pupil: 'Lessons With Plants' was made to supplement the work of the teacher." This is the opening sentence of the preface, showing that the book is a companion to "Lessons With Plants," which has now become a standard teacher's book. The present book is the handsomest elementary botanical text-book yet made. The illustrations illustrate. They are artistic. The old formal and unnatural Botany is being rapidly outgrown. The book disparages mere laboratory work of the old kind: the pupil is taught to see things as they grow and behave. The pupil who goes through this book will understand the meaning of the plants which he sees day by day. It is a revolt from the dry-as-dust teaching of botany. It cares little for science for science' sake, but its point of view is nature-study in its best sense. The book is divided into four parts, any or all of which may be used in the school: the plant itself; the plant in its environment; histology, or the minute structure of plants; the kinds of plants (with a key, and descriptions of 300 common species). The introduction contains advice to teachers. The book is brand new from start to finish.

"An exceedingly attractive text-book."—*Educational Review*.

"It is a school book of the modern methods."—*The Dial*.

"It would be hard to find a better manual for schools or for individual use."—*The Outlook*.

THE MACMILLAN COMPANY

64-66 Fifth Avenue   NEW YORK

THE CYCLOPEDIA OF AMERICAN HORTICULTURE: By L. H. BAILEY, of Cornell University, assisted by WILHELM MILLER, and many expert cultivators and botanists.

4 VOLS.— OVER 2800 ORIGINAL ENGRAVINGS — CLOTH — OCTAVO $20.00 NET PER SET. HALF MOROCCO, $32.00 NET PER SET

This great work comprises directions for the cultivation of horticultural crops and original descriptions of all the species of fruits, vegetables, flowers and ornamental plants known to be in the market in the United States and Canada. "It has the unique distinction of presenting for the first time, in a carefully arranged and perfectly accessible form, the best knowledge of the best specialists in America upon gardening, fruit-growing, vegetable culture, forestry, and the like, as well as exact botanical information. . . . The contributors are eminent cultivators or specialists, and the arrangement is very systematic, clear and convenient for ready reference."

"We have here a work which every ambitious gardener will wish to place on his shelf beside his Nicholson and his Loudon, and for such users of it a too advanced nomenclature would have been confusing to the last degree. With the safe names here given, there is little liability to serious perplexity. There is a growing impatience with much of the controversy concerning revision of names of organisms, whether of plants or animals. Those investigators who are busied with the ecological aspects of organisms, and also those who are chiefly concerned with the application of plants to the arts of agriculture, horticulture, and so on, care for the names of organisms under examination only so far as these aid in recognition and identification. To introduce unnecessary confusion is a serious blunder. Professor Bailey has avoided the risk of confusion. In short, in range, treatment and editing, the Cyclopedia appears to be emphatically useful ; . . . a work worthy of ranking by the side of the Century Dictionary."—*The Nation.*

This work is sold only by subscription, and terms and further information may be had of the publishers.

THE MACMILLAN COMPANY

64 66 Fifth Avenue   NEW YORK